We, Robot

NEDO

Also See Other Books By Mark Stephen Meadows:

Pause & Effect: The Art of Interactive Narrative
I, Avatar: The Culture and Consequences of Having a Second Life
Tea Time with Terrorists: A Motorcycle Journey Into the Heart of Sri Lanka's Civil War

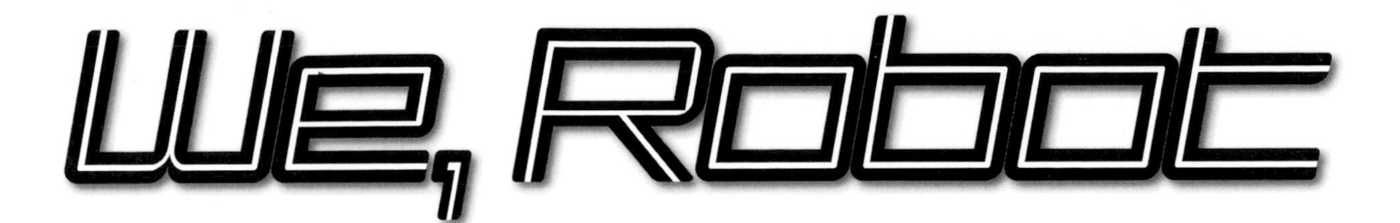

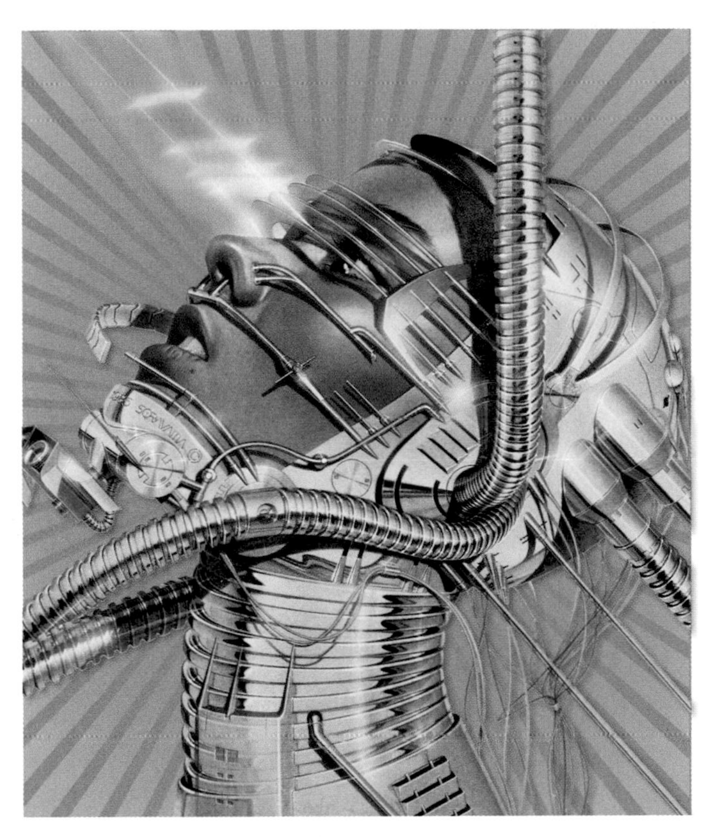

Skywalker's Hand, Blade Runners, Iron Man, Slutbots, and How Fiction Became Fact

Mark Stephen Meadows

LYONS PRESS
Guilford, Connecticut
An imprint of Globe Pequot Press

Lyons Press is an imprint of Globe Pequot Press.

Library of Congress Cataloging-in-Publication Data is available on file.

ISBN 978-1-59921-943-1

LYONS PRESS

Executive Editorial Director	Janice Goldklang
Acquisitions Editor	Keith Wallman
Project Editor	Kristen Mellitt
Design Manager	Elizabeth Kingsbury
Book Designer	Bret Kerr
Production Director	Lisa Nanamaker
Layout Artist	Kevin Mak
Manufacturing Director	Kevin Lynch

Printed in the United States of America

10 9 8 7 6 5 4 3 2 1

Contents

PART TWO: FUTURAMA

List of Illustrations

List of Illustrations

Introduction: The Virginia Calculator

Shutterstock

IN THE FALL OF 1789, TWO WELL-TO-DO GENTLEmen named Samuel Coates and William Hartshorne were riding in a horse-drawn carriage through the Virginia countryside. They were in search of a calculator. Of course, it being 1789, calculators were pretty rare, and these two had traveled some distance to find this one, which had a reputation of being the finest in the land. The calculator they sought was so fast and so respected among the people of Virginia that it had become something of the pride of the state.

Mr. Hartshorne and Mr. Coates had spent many months researching calculators (there were several in the United States and Europe at this time), and they had contacted the owner weeks in advance by formal post, with a request to see the calculator in operation to best certify the rumors of its speed and accuracy. Elizabeth Coxe, of Alexandria, was the owner. She came from a wealthy family, able to afford such luxuries and objects of interest, and our two researchers were surely on their best behavior when their carriage arrived at her estate.

We can assume that after appropriate introductions and niceties were completed, the work began. The calculator was brought in and the group began solving several equations, some of them relating to the number of seconds there are in a year and a half, and another question relating to geometric calculations of land size.

Mr. Coates, pen in hand, scribbled as fast as he could, trying to confirm the precision of the numbers, but pen and paper proved to be much slower than the calculator. More numbers were piled on, more calculations were performed, and the calculator just kept cranking out the answers with unbelievable speed and precision.

On the final question a flaw appeared; the calculator made a mistake on this, the most difficult question of the session. The problem asked how many seconds a man has lived if he has been on the planet seventy years, seventeen days, and twelve hours. The calculator produced its number and Mr. Coates, some minutes later, caught up with his own. The numbers did not match. Coates held up the paper with one hand and, pointing at the calculator, indicated that while fast, it was wrong.

Our cultural perceptions and morality are due for another reboot.

The calculator said, "Stop, master! You forget the leap year!"

Mr. Hartshorne and Mrs. Coxe looked first at the calculator, then at Mr. Coates, who promptly replied by adding this factor into his own calculations. After a few moments' work, the numbers matched.

The Virginia Calculator was named Thomas Fuller (1710–1790). He had been brought to Virginia and sold as a slave when he was fourteen or fifteen years old. Since he had not been captured until his early teens, he had reaped the benefits that many people from the Ivory Coast had gained—Arabic calculation abilities that were employed in African markets for hundreds of years. Mr. Fuller had the knowledge of this math in addition to a natural talent for visualizing large number sets.

Abolitionists pointed to Mr. Fuller as proof that blacks were intellectually equal to whites. Although the Emancipation Proclamation wouldn't be signed until 80 years later, in 1863, Mr. Fuller was a beacon for events to come—not just the emancipation of slaves in 1863, but an age dawning today, in the early part of the twenty-first century.

In the coming decade, just as in the 1600s and 1700s, slaves with the ability to calculate will be bought and sold. These slaves will symbolize their owners' social standing, and they will perform for visitors that have traveled to see them. We will not think of them as our equals. We will think of them as our property. These slaves will be called robots.

These new slaves may be machines rather than men, but, just as happened in the 1700s, our cultural perceptions and morality are due for another reboot.

Shutterstock

Automated manufacturing: the start of it all.

Eventually, of course, machines would be able to calculate much faster than Mr. Fuller; they would be able to work faster than humans in many ways, replacing men, women, and children in all kinds of industries, from agriculture to textiles and beyond. Some people gained their freedom and some people lost their jobs.

In 1955, Detroit's "Big Three" automakers controlled the majority of the world's automobile production resources. Almost all of this was due to the 500,000 employees of General Motors—a company so powerful that its only concern was an antitrust suit from the U.S. federal government. But now, more than fifty years later, things look different. Most of GM's

Mark Stephen Meadows

The Wowee RS Companion Tri-bot.

work has gone to Japan, a comparatively small country, with comparatively less wealth and fewer people. But Japan is a country that relies heavily on robots. These robots work for less, produce more, don't need health care, are able to work longer hours, and they don't organize inconvenient unions. In fact, the majority of the world's robots operate in Japan. According to the United Nations Economic Commission, there are over 500,000 robots operating in Japan—the same number of people that worked for General Motors in 1955.

In the automotive industry most robots pay for themselves in about eighteen months. The one-time cost of a robot is about as expensive as an employee's salary for two years. For example, in 1984, the cost of an automotive industry welding robot was about 12 million yen, and a worker's salary was 5 million yen. And, as the cost of robotics decreases, and the quality and quantity of the work the robot can do increases, robots will become even more valuable than they are now. Robots will replace more people and will do more work. From a financial perspective, this means that robots will become more valuable than humans.

The automotive industry isn't the only industry whose employees are being replaced by machines. Calculators and computers used to be real humans that sat at real desks and worked with real pencils and paper. Cotton pickers, corn huskers, loomers, textile workers, plumbers, masons, canners, bottlers, glass makers, bread bakers, car washers, and surgeons have all lost their jobs to robots. In the 1990s the trend toward automation moved increasingly into informational and cultural areas. Bank tellers, telephone operators, customer service representatives, actors, designers, and musicians—even the guys in the underground labs in Los Alamos, New Mexico, who tell the rockets where to fly, or the guys in New York offices who tell the stock market where to fall—all of these people have been replaced by robots.

If the 1980s was the decade when corporate mainframes became personal computers, the 2010s will be the decade when industrial robots will become personal robots. Today we can find robots that vacuum floors, trim lawns, clean windows, cook pancakes, and fold clothes; we can find some that are pets, and some that go to war. Robots exist that do all of the dirty, dangerous, or deadly work that we humans don't, or can't, do ourselves. Robots are performing more of the automated and rote labor that humans quickly tire of; best of all, robots cost less and ask for nothing in return.

Robots will become more valuable than humans.

For a while it seemed we had nothing to worry about. Robots used to be old-timey science fiction conceits—clunkers that carried bikini-clad women, fired lasers, and lived on Mars or the Moon. They existed far into the future. Sometimes they looked like us, as in *Blade Runner*'s Roy, or Pris, but those robots lived in the future, too. And although *The Terminator* looked a bit like us, that movie came out way back in 1984, so it wasn't a worry.

Then, about ten years ago, things started getting weird. Robots started walking—real robots, like ASIMO, really walking. And then there were robot pets in people's houses, like Aibo, and robot vacuum cleaners named Roomba started scootering around our living rooms, collecting crumbs.

Recently things seem to be getting crazy. Robots are all over the place. I've heard about a robot in San Diego that can smile at itself in the mirror, and another in Tokyo that carried an elderly woman to a waiting ambulance. I've heard about robots that can climb stairs and walls, and one that can climb inside your heart, like a mechanical worm, to perform surgery. I've heard about robots launched to Afghanistan (yet controlled from Nevada), and robots launched to Mars (yet controlled from Earth). In 2010, I heard about a robot in the online world of Second Life that was selling virtual sex, while another in the gaming world of Las Vegas was selling real sex. Well . . . real *robotic* sex.

We now have robots operating as soldiers, firefighters, train conductors, fruit pickers, shoppers, sellers, and ticket vendors. We have functioning robotic car valets, painters, lawn mowers, floor scrubbers, and chefs, and robots that will efficiently stow your dishes after they've been washed and pulled from your fancy newfangled robotic dishwasher. That's

Shutterstock

The difference between robots and humans is getting thin.

if you're feeling lazy, because we also have robots that can interpret your mood from your gestures or your tone of voice. And we even have little seal-shaped robots that have become quite popular as therapists.[1] If you pet it, the thing will look you in the face with its big, shiny, baby-seal eyes, squirm a bit, and make a creepy *meow* sound.

There are now virtual robots, cyborgs, fembots, chatbots, slutbots, androids, gynoids, geminoids, warbots, and hybrots. In 2009 one robot was showing signs of the swine flu (H1N1) virus in Tokyo, while another was watching for woodpeckers in Arkansas. We can control robots from a distance now, with just a thought (and a rather bulky helmet). It's called a "brain-robot interface." So a robot is now a kind of avatar. And there are prosthetics that are called robots that help crippled people walk, or that assist rescue workers in saving people after a disaster.

WTF? Are robots out-evolving us? Or is this just the first clatter and rattle of our future overlords as they slowly rise to their caster-roller feet? Is this stuff dangerous? Are we building something that will unbuild us?

[1] (In a reversal from Asimov's tale *I, Robot.*)

What is a robot, really? What does this word even mean?

I went to Tokyo to find out.

When I got there I assumed I knew what a robot was, but assumptions are those things that happen when you're not paying attention. When you find yourself standing in front of a silicone cadaver as its eyes follow you around the room, you start to question them—the assumptions, I mean, not the robots. For a while, I had assumed that a robot was something that replaced a human; then I thought it was something that did what a human could do, only better; then, I read an essay by Bruno Latour[2] and got really confused about what a robot is.

In Tokyo, I had the chance to ask the top people in the field what they thought. After visiting three cities and more than two dozen research labs and companies, seeing nearly 150 robots and talking with dozens of inventors, I am relieved to report that robots today are more puppets than people.[3] I discovered that robots aren't as advanced as I'd imagined. I found big plastic avatars; I didn't find artificial intelligence, and I never found anything resembling SkyNet, the fictional computer network in the *Terminator* series. There is no great artificial intelligence that I am aware of—at least, not in the foreseeable future. Robots are, and will continue to be, guided by humans, and humans will continue to make the key decisions.

Robots, ultimately, are us—and while this is somewhat of a relief to report, that news should also launch a concern or two. We're becoming robotic ourselves, as robotic technology is being built into human bodies: eyes, arms, legs, hearts, and ears. Mentally, too, robots are helping us make decisions on finances, politics, and family. Robots and brains can now be connected, chemically and electrically. So both in terms of body and mind, we're becoming more like robots as robots become more like us.

Today there are nearly nine million robots in operation.[4] The Japanese Robot Association predicts that by 2021, the personal robot industry will increase to more than $35 billion a year worldwide, compared with about $6 billion today. The South Korean Ministry of Information plans on having a personal robot in each South Korean household by 2020.

During my travels to Tokyo I decided that a robot is something that is part myth, part tool. It is a tool that has taken on mythical proportions. Although this book will offer definitions, it will also provide some context in terms of the changes around us, some information on what

[2] "Where are the Missing Masses? Sociology of a Door."

[3] Even the International Organization for Standardization (ISO 8373) definition of a robot—"an automatically controlled, reprogrammable, multipurpose, manipulator programmable in three or more axes, which may be either fixed in place or mobile for use in industrial automation applications"—can be considered a definition of a computer.

[4] Source: International Federation of Robotics, IEEE, and World Robotics Research, Executive Summary of World Robotics 2010 report, www.worldrobotics.org.

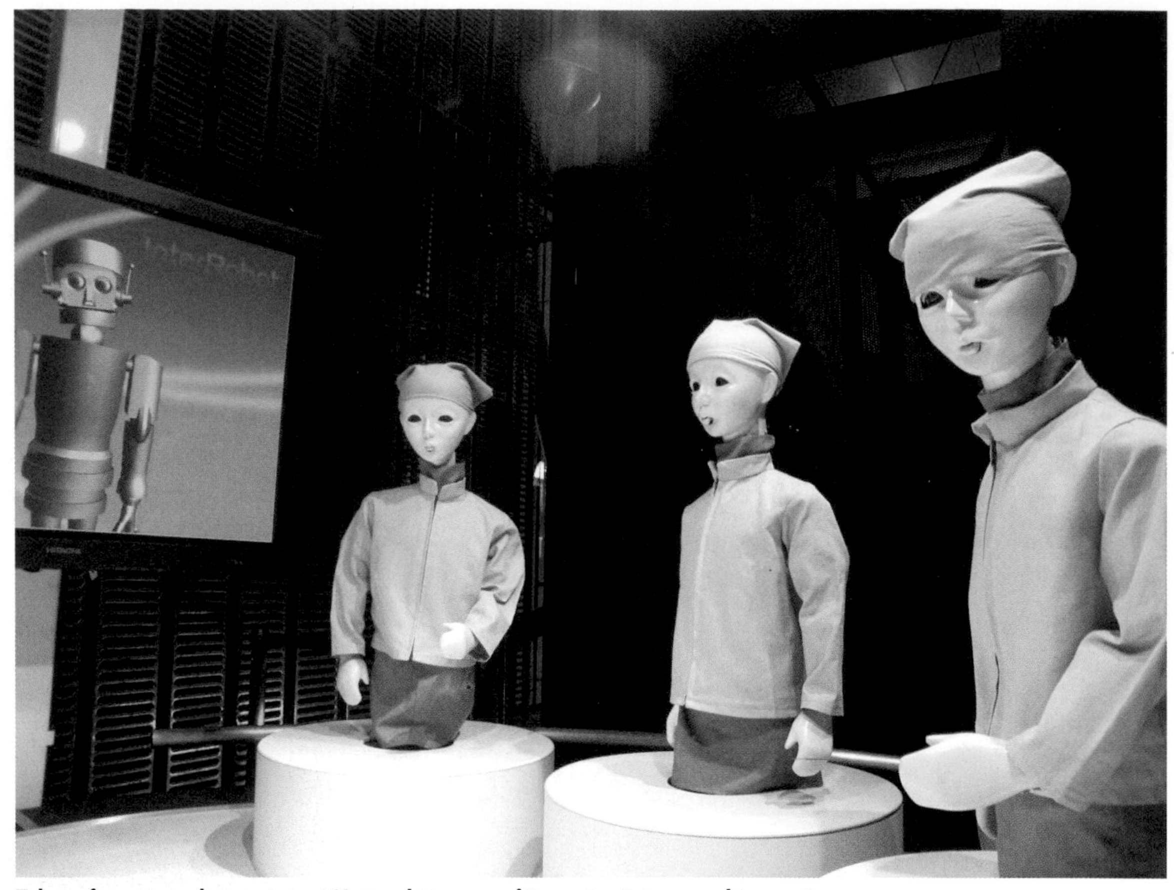
Mark Stephen Meadows

Teleconferencing robot-avatars at National Museum of Emerging Science and Innovation.

to watch out for, and most of all, a reminder that we are our robots and they are us.

We, Robot will introduce you to the latest discoveries in robotics and artificial intelligence.[5] By looking at science fiction universes (*Blade Runner, Star Wars, Battlestar Galactica,* etc.) and robot residents (Iron Man, Terminator, and HAL 9000, to name a few), the book shows how close they are to becoming a reality, and, in some places, where reality has already rocketed past. Through a discussion of robots loved and hated by millions, it provides an informed, entertaining, and often surprising look at what to expect in the next year, as well as the next decade. The book is structured in two parallel parts (today, and the coming decade), with four chapters on emotion, interaction, intelligence, and body.[6]

[5] We will also dust off some old topics that have been around since the 1950s, such as Marshall McLuhan, Sherry Turkle, and others.

[6] That is to say that chapters 1 and 5 are both on emotion, chapters 2 and 6 are both on interaction, etc.

We, Robot is about how, if you take away the robot's faceplate, you will find a human behind it, and if you tweak that human nose, you might find it to be surprisingly metallic under the pliable skin.

It's as if we're building a new slave class.

It's as if we're building a new slave class. Millions of new Mr. Fullers will soon be appearing. Will another 1863 appear, when this slave class will demand equal rights? Will they want burial ceremonies? Will we marry them or have sex with them? Will we be able to tell them apart from us?

We, Robot looks at these questions (and irresponsibly raises more), offering some introductions to the people and robots in the middle of the field today. In this book I stretch the word *robot*, discover some strange creatures, get lost a few times, and get found at least once. The book is not a series of points to be proven, but a number of stations to be visited, and although I make predictions, I don't know what will happen—they are just my own science fictions. Throughout, there is one certainty: We will all be affected by robots. And that is part of what makes this topic so interesting.

Our story begins in the magic robot kingdom . . .

Part One

Today

Mark Stephen Meadows

Chapter One: The Terminator

On Robots That Hunt Humans, Have Faces, and Why We Are So Afraid of Them—Emotion, Part 1

> Hello? . . . Ha ha. Fooled you. You're talking to a machine.
> But don't be shy, machines need love too.
> —Sarah Connor's Telephone Answering Machine,
> *The Terminator*, 1984

I STEP OFF MY TRAIN AND ONTO THE PLATFORM of Tokyo's Shinjuku metro station. I'm standing in the middle of the future.

This station is a shopping mall labyrinth, a Borgesian library, an urban implosion of neon signs, tow-along luggage, vertical writing, bustling and running people, handheld computers, and me, an utter illiterate, in the middle of it all. I can only murmur bad pronunciations of the Japanese words for *thank you, hello,* and *excuse me*. North and south have abandoned me, there's no sun to be seen, and I can barely tell up from down—and not because I'm jetlagged. It's because I'm in Tokyo.

As I stand there on the wrong side of the metro tracks, trying to figure out which part of my map is north, an old man with a shiny blue suitcase and shiny blue eyes asks me if I need help, in "Engrish," and I eagerly accept, thanking him. I pride myself on my ability to get around, but this is the first time in some six or seven years that I haven't been able to read a letter, much less pronounce a word. He gives me a hand with directions and I give him a hand with his suitcase, and five minutes later, I finally make it out of the airports and train stations that have been dominating my life for the last several days.

Up the escalator, squeezing in between people, past noodle shops and vending machines, under phosphorescent lights, through gate control, and out onto a large esplanade at the Shinjuku-Nishiguchi Station, into the fresh night air where winter hits my face like a bag of ice. There's some more ramen shops, a McDonald's next to them, strange pointed buildings in the twilit distance, lights laughing everywhere, writing covering everything, strange tiny characters of an alien grammar dancing on billboards and signs, signs, signs, all of them civic screens and public computer monitors. People are beautifully dressed, very clean. They are as stylish and solid as the guardian buildings that protect the street from the world beyond.

I, Robot Hunter

TOKYO IS BIG AND MOVING FAST. I DEFINITELY need a taxi right now, and there are a few lined up along the esplanade. I dig out my directions, approach the taxi that's waiting, lean down as I walk up (partly to see him, partly to let him know I'm there), and reach for the door. The thing pops open and bangs my knuckles. I recoil, thinking he's opened it from the inside. But no, the damned door keeps opening and pushes against my knee, and I take a second step back so as not to get pushed over. It's automatic and I remember, too late, that a friend had warned me of this little automated surprise.

I'm anxious to meet my first cyborg chauffeur.

All hopped up on my robot hunting, I slide into the cab, anxious to meet my first cyborg chauffeur. Will he have a silver pate? Will he have one red eye?

No, of course not. A normal human turns around and smiles at me. But his warm expression rapidly turns to one of confusion as I try to show him the whereabouts of my hotel on my already-rumpled map.

We pull away from the curb. Not a word has been spoken (at least, not one that both of us could understand), but we get the basic idea across, and off we go. The lights slide past my window and the back of the seat in front of me changes colors, lit by the passing lights that wash past us as we drive to the hotel.

Looking out the window I wonder about the definition of a robot, and the lines we draw between ourselves and our bodies, our bodies and our machines. I look up at the buildings and the dense neon signs as we cruise through the noisy streets, filled with people. There are rows and rows of vending machines lining the streets. I see a huge Gundam robot at a street corner, lifting its hand slowly in the air, and I watch another taxi, as we pass it, opening its door for a passenger who's just getting out.

As the lights wash by, I remember a short film clip I saw a few months ago. It was a product demonstration about a robot, and it seemed believable. It was a four-legged contraption named "BigDog,"[1] and it looked like a cylinder mounted on four legs, but with one set of knees reversed. It was a headless body with mechanical legs, really, a simple pack mule with complex algorithms. It had a rather uncanny way of walking. It ran like a drunk deer, but what happened next proved its sobriety. A guy in the video walked along next to the robot and kicked it, and pretty hard. Although the robot

[1] You can find it on YouTube, or Google it by searching for "big dog boston robotics."

Mark Stephen Meadows

Tokyo at night.

was pushed to the side, it deftly regained its balance. Even with some seriously slippery conditions, this algorithmic pack mule had a good enough sense of balance to stay upright. Not only could it regain its balance while walking on ice, but it also knew where to go, what rocks to avoid, where to step, and appeared capable of jumping over premarked barriers on test treadmills.

Pretty good for not having a head. It was science fiction turned engineering fact, like Tokyo—a real-life science fiction city. A robot that had made its way backwards from the future. It was believable.

But *The Matrix* was believable, too. In fact, it was so believable that some respected researchers and scientists have calculated the probability of our actually living in a *Matrix*-like simulated reality.[2] Probably on someone's computer in the future, they told us.

The film *2001: A Space Odyssey* seemed believable, too. *Battlestar Galactica, Blade Runner*, or any one of a number of other films that take place far enough in the future to be fictional all take on a kind of believability. My disbelief is so suspended these days that I'm floating into the future, totally unmoored from any sense of what can and can't be done. News

[2] This argument is commonly called "The Simulation Argument," and more can be found at www.simulation-argument.com.

I read just supports the credibility of robotic advancements. I recently watched a video of a monkey that had wires connected between his brain and a mechanical arm, and the little guy was able to feed himself a banana with it. He was probably pretty hungry after the surgery. And I've read that we are building bionic limbs and bionic eyes, and are moving beyond simple cochlear implants; the price of these bionic improvements is actually less than $6 million. Cyborgs? Yep, in a recent issue of *National Geographic*.[3] I just don't know what *not* to believe anymore. What's equally disturbing, and probably related, is that I'm rarely surprised by new advances I learn of, especially in the field of robotics.

The very concept of a robot originated in science fiction.

It all seems like the science fiction that I've read and watched over the years.

I came to Tokyo to find out what the word *robot* means. I came here to interview robots and their makers, to learn about the cultural contexts and mentalities. Along the way I've also visited Los Angeles and Paris, and spoken with nearly a hundred people in over twenty countries, so that eventually I might understand this mythic moniker, *robot*.

The very concept of a robot originated in science fiction. The word *robot* comes from the Czech word *robota,* which roughly means "indentured work."[4]

In 1920 a Czech author by the name of Karel Čapek finished a play titled *R.U.R.* (or *Rossum's Universal Robots).*[5] The play takes place in a factory where robots—androids, or human-shaped robots, specifically—rise up against their creators. The play was no great success in Europe. Despite the tepid reception, or perhaps because of it, a few versions of the play made their way across the Atlantic, and though they didn't make it to Broadway, they did make it as far as New York and into the hands of a twenty-eight-year-old man who had just received his PhD in biochemistry. His name was Isaac Asimov.

Dr. Asimov was getting to work on a series of short stories that would end up being compiled in the book *I, Robot.*[6] With the book's 1950 publication, the word *robot* slipped into common English, and with it, the concept of the field of robotics, as well. Shortly after that came the

3 January 2010, "A Better Life with Bionics."
4 The kind of work you might be obliged to do for a lord or sovereign.
5 http://www.gutenberg.org/etext/13083.
6 Asimov argued for the title of the book to be *Mind and Steel,* but his editor wouldn't go for it.

idea of a robot psychologist. While the American public was watching black-and-white television and congratulating themselves on the purchase of a toaster, Isaac Asimov was setting robots up in therapy sessions and considering when—and how—they would rule the world.

In both *R.U.R.* and *I, Robot,* the stories take place in the not-too-distant future, in which technology has become so human that the very people who created it can't make out the difference between the robots and the humans. In each of these stories there comes a moment when the robots and the humans are indistinguishable; a test has to be given, and shortly afterwards the robots rise up and take control.

I guess we're near that point in real history now.

There are other similarities between the two works. In both, the robots are brewed with a combination of genetic engineering and electrical engineering. They become a global industry because they cut costs and increase profits. They have emotions and fall in love. Different kinds of robots have different functions, and one of them catches hold in a sort of artificial evolution. An authority conflict breaks out and fisticuffs take place. The story ends with robots taking over the world.

So I guess we're near that point in real history, now, too?

Genetic engineering, like *robot,* is another term that sprang from science fiction,[7] along with *utopia,*[8] *computer virus,*[9] and *cyberspace.*[10] The latter word has now been adopted by the boys in Washington, D.C., as they set up "cybersecurity" features at the Pentagon.

In fact, science fiction may be less a prediction of the future than a warning about the present. Science fiction, like almost all forms of art and literature, is a bellwether and weak-signal sensor of current cultural trends. It has been a means of shining a light on cultural and ethical issues, a safe space to discuss dodgy topics. Science fiction might be an early warning system that uses the minds of artists and authors like sensitive earthquake prediction centers. The genre's authors comment about our present more than they predict our future. For example, in 1948, when *R.U.R.* was written, slavery had just recently been abolished in Czechoslovakia by the League of Nations' Slavery Commission. Čapek not only saw these changes during his lifetime, but he had, himself, worked toward affecting them. So his play is about slavery, and the word *robot* was chosen to point that out.

District 9, another great science fiction film, takes place in Johannesburg and is about racism, and the director, Neill Blomkamp, has flatly said as much.[11] And James Cameron, the

[7] According to Bruce Sterling in an article for *Wired* magazine, the term came from a book titled *Dragon's Island* by Jack Williamson.

[8] "On the Best State of a Republic," Sir Thomas More, 1516.

[9] Again, via Sterling, Dave Gerrold used the term as it refers to computer software in a 1972 story, "When Harlie Was One."

[10] William Gibson coined the phrase in 1982 and 1984 with "Burning Chrome" and *Neuromancer,* respectively.

[11] *LA Times* interview with Geoff Boucher, January 2010.

director of *Terminator, Avatar,* and other science fiction classics, is quoted as having said:

> *... when I wrote the first* Terminator *outline around 1982, I was just working out my childhood stuff. It was also born out of the science fiction movies and literature I grew up with. For the most part, they were warnings—about technology, about science, about the military and the government....*[12]

By the time I get to my hotel room it's about 9 p.m. (10 a.m. my local time), and I'm not even thinking of sleep. I spend the evening soaking up packets of instant coffee, researching robots in Japan, following up on e-mails I'd sent to inventors, entrepreneurs, research labs, and corporations, while listening to the television in the next room emit muffled screams. The hotel room smells of cigarettes.

The night rolls into dawn and I'm hungry. Finally, trembling and unable to type from all the instant coffee, I decide to take a break. I pull on my jacket and head out into the street. I shuffle up the empty sidewalk a couple blocks, find a little all-night convenience store, and, unable to say more than "Hello" to the nice man at the counter, I point to some fried lumps of something-or-other in the window. He puts them in a little folded piece of paper for me and patiently waits as I also select some toothpaste and a bottle of water.

Stumbling back home, I sit down at the desk, eat what seem to be chicken nuggets, throw down a couple of codeine for dessert, drink some more coffee for the hell of it, and use the hotel room's high-bandwidth connection to drag the sun up into the morning sky. My first appointment is in three days. My toothpaste turns out to be hand lotion (which I discover in the worst way imaginable), but it's all just fine with me because I'm in Tokyo.

Robots Gone Wild!

ONCE UPON A TIME,[13] SOME NEC CORPORATION RESEARCHERS equipped a PaPeRo model robot with an infrared spectrometer that analyzed wines. Mounted on the robot, as a kind of a sensing arm, the little guy could figure out the composition of food and drinks. This was done by shooting a beam of light into a bottle and, a bit like a bat senses an echo and determines distance, the colors that bounced back detailed the chemical composition of the wine. Mid-infrared and near-infrared rays could determine molecular structures, which meant that, as a method of determining chemical makeup, and, hence, the wine's flavor, you never had to open the bottle (which is definitely not why I, myself, buy a bottle of wine, but it's probably fine for sommeliers and auctioneers).

So imagine that you have your wine cave and your robot, and you set him down on the ground and point to a bottle. The robot would

12 http://www.wired.com/entertainment/hollywood/magazine/17-04/ff_cameron.

13 Expo AICHI, June 2005, in Japan.

Mark Stephen Meadows

PaPeRo. Cute, and can smell bacon on your hand.

roll up to the shelf, point its little arm at the bottle, pause, and then say, "*Beep.* Red fruit sweetened with chocolate. Clove. Cinnamon. Black cherry, slight licorice, butter with finish of vanilla. *Beep!* Probably a 1998 Pinot Noir from the Bourgogne region."

The robot could also make recommendations. PaPeRo was not only able to peg wine composition; it was also able to determine which apple of three would be the most sour, which cheese of three would be the most salty, and could even be programmed to warn its owner, in its monotone and rather childlike voice, which foods contained high quantities of fat, salt, or carbohydrates.

Something will surely run amok.

So NEC would do presentations of this robot, largely for the press, and during one of these demonstrations, a smart-ass photographer had the idea to put his hand in front of the PaPeRo's sensor. The robot paused, analyzed the hand, and declared it to be bacon. Of course, this caused horrified laughter among the group, but in case there was an error, and since one smart-ass inspires his fellow, a nearby reporter then tried the same trick. His hand was identified as prosciutto, another pork product. Of course, this again shocked and elated everyone and made lots of press around the world, despite its being slightly different from the kind of attention NEC had hoped to receive.

One can't do much more than giggle, since, after all, it's one thing for a robot to smell flesh, but it's altogether another for said robot to "eat" it, especially for little green-and-white PaPeRo, which is locked up in a research lab and only trotted out for the occasional press conference (when he's not doing his normal job, which is taking care of children).

We can reassure ourselves that PaPeRo is no anthrovore—it's simply a sensor that helps humans sample molecules from a distance. Robots do not eat, especially meat. They are plugged into walls and run on battery power.

But *wait.* Not necessarily. Robots need amperage to move servos, just as we need calories to move muscles, and it is possible to squeeze electricity from almost any organic material. It's called a microbial fuel cell.

About twenty years ago it was discovered that enzymes, when they gobble up almost any organic matter, can produce electrons. This amounts to a bioelectrochemical system that generates electricity by using bacterial interactions in some rather queer conditions. It's a great way for a city to generate power from the processing of wastewater, for example, or for getting a car to run on compost. Or for feeding a robot.

There are now robots in existence that sit in the living room and are capable of eating mice, or flies, and extracting electrical energy via microbial fuel cells. In fact, I asked one of the inventors of these systems, James Auger, how much power 4 or 5 kilograms of flesh would generate. That's about what a human arm would weigh.[14] Mr. Auger wasn't keen on answering this question, as he prefers a more-subtle messaging approach (he's English, after all), but what was clear is that these robots can go a long time on a little biomass.

Well, that got my hyperactive imagination going: Beware *C.H.U.D.*, beware mole people, beware oh, ye alligators of New York.

Ah, the old science fiction nightmare: Something will surely run amok. Some misanthropic, overambitious robot designer in some deep, dark underground lab will set loose a murderous, flesh-eating invention that will kill its inventor, smash through the nearest wall, and begin its relentless rampage against the outside world. The headlines the next morning will read FLESH-EATING ROBOT ESCAPES LAB! The accompanying photo will show the genius inventor, still dressed in his white lab coat, slumped next to the gaping hole in the wall. Part of his hand will be missing, and somewhere in the caption will be the words "Dr. Frankenstein."

Here in the West,[15] this fiction is the meat and potatoes of robot films. It's as much a part of science fiction as the science part. The robot uprising is now its own literary genre. In fact, it was such a common occurrence that Isaac Asimov himself decided at one point to try to change this; in *I, Robot* (and, more specifically, in the short story, "Runaround"), he famously did so.

Asimov decided, once and for all, to propose a solution to this renegade robot problem. It would be something hardwired into the robot's positronic[16] brain. It would be a means of assuring that robots would behave themselves. The three laws are like robot instinct:

1. *A robot may not injure a human being, or, through inaction, allow a human being to come to harm.*
2. *A robot must obey any orders given to it by human beings, except where such orders would conflict with the First Law.*
3. *A robot must protect its own existence as long as such protection does not conflict with the First or Second Laws.*

These moral algorithms, Asimov famously pronounced, would protect us from our own inventions (once the robots got smart enough to figure out something was wrong—such as the fact that they'd been enslaved, say). These instruction sets for ethical programming[17] would

[14] A human arm is commonly about 6.5 percent of a modern human's body weight.

[15] Curiously, this does not seem to exist in Japan. Robots began with Astro Boy rather than *R.U.R.*, and so the foundation of the robot myth includes a cute little fellow that helped out, not a colony of killer clones. Kudos to the Japanese for preserving this legend as it is, today.

[16] *Positronic* was, at the time, another candidate for what we know today as electronic (positrons vs. electrons).

[17] Asimov was more interested in ethics than robots.

protect us from The Robot Uprising. Asimov's three laws would prevent groups, such as the military, from creating a robot that would break loose, cut all of us *long pigs* into strips of bacon, and eat us all for breakfast.

But in the world of engineering fact, we're not being cut up for bacon, nor are robots being designed with Asimov's three laws. This is because our cultures, psychologies, and reasons for building robots in the first place are not guided by reason, or design, or some evolving super-mind that is out of control.

The technology isn't fearsome—the people that build it are fearsome.

It's the Dr. Frankensteins that get us in trouble.

How the Autocalypse Goes Down

AS YOU PROBABLY KNOW, *THE TERMINATOR*, JAMES Cameron's 1984 film, is a science fiction classic starring Arnold Schwarzenegger, Linda Hamilton, and Michael Biehn. It grossed over $500 million worldwide.[18] It is one of the hundred most popular films ever made. It eventually even hit television series status as a cult favorite titled *Terminator: The Sarah Connor Chronicles,* which aired for two seasons on Fox Broadcasting. Each iteration became weirder until, last year, when I saw *Terminator Salvation,* I couldn't keep track of why there were so many explosions. But the first movie kicked unadulterated ass, and it left me with psychological scars that I still fondle to this day.

The original movie, which presents the true core of this science fiction vision, takes place in 1984, when a "Terminator" (a cyborg from the year 2029) has traveled backwards in time to carry out an assassination attempt. The robot is, technically, a *cyborg,*[19] a robot-human hybrid, and this one has been sent by its colleagues, a group of superintelligent machines, hell-bent on the extermination of the human race, to kill Sarah Connor, the soon-to-be mother of a key revolutionary. Meanwhile, in this distant future, the key revolutionary is aware of this plot to kill his mom, so he sends Kyle Reese, a human, back in time to kill the Terminator before it can kill Sarah.

Now, the Terminator is a robotic assassin. It's a soldier designed for infiltration and combat duty. It can speak and mimic the voices[20] of people it has heard. It has bad breath and can bleed, sweat, drive a motorcycle, and perform minor surgery on itself. Otherwise (unless you happen to be a dog, especially a German shepherd), it is indistinguishable from humans. It has "Strong AI" (artificial intelligence) and eyes that glow red. The earliest Terminator was the Cyberdyne systems Model 101, specifically the 800 series. With a "hyper-alloy combat chassis" in the form of an endoskeleton, it

[18] http://boxofficemojo.com/showdowns/chart/?id=vs-terminator.htm.

[19] If we get more technical, the Terminator is an android cyborg.

[20] As well as personalities, it seems.

is controlled by a microprocessor, is covered in living tissue, and looks almost exactly like Arnold Schwarzenegger.

According to what we know from the movie, the Terminator is the tough little nephew of a failed military experiment named SkyNet. This first Terminator is finally terminated in Sunnyvale, California, in the heart of Silicon Valley. It happens on the manufacturing floor of the company called Cyberdyne, which also happens to be a robotics workshop.[21] This means that after Sarah is hauled off in the ambulance at the end of the movie, the engineers of the shop open up the freshly crushed Terminator and reverse-engineer the thing. Thus, the Terminator is born. Cyberdyne sells the tech to the U.S. government, which sells it to the air force, which then launches a project called SkyNet.

This is where things start to go wrong. SkyNet is a product of this division of the air force, Cyber Research Systems (here's where Frankenstein comes into the story), which is producing a network of AI systems intended to replace military and civilian aircraft pilots. SkyNet is also intended to control military weapons systems, including nuclear missiles (so the problem starts in trying to replace people, and in trying to develop autonomous weaponry).

According to *Terminator* legend, the system went online on August 4, 1997, and twenty-five days later, it became self-aware. SkyNet's human operators try to shut it down, but in an act of self-preservation and retaliation, it launches a nuclear attack against Russia, who retaliates against the United States. A global thermonuclear war breaks out, the humans are reduced to rats, the machines get busy, and the Terminator finds his proper place in the world.

The technology did not get out of hand because of the technology.

The technology did not get out of hand because of the technology. The technology got out of hand because of the group of people in the U.S. military. This isn't the trend in just *The Terminator*. It's the same thing that happens in *I, Robot*, and other stories, as well.[22]

Science fiction authors everywhere love this theme more than any other. Robots won't turn us into bacon bits to power their myoelectric power supplies, and SkyNet didn't become self-aware in 1997. But science fiction is able to present possible futures. The word *robot* is one, and military robotics is another. But are they legitimate concerns?

Soldiers Gone Wild!

WHEN THE TERMINATOR SAID "I'LL BE BACK," HE wasn't kidding around. In fact, he's been

[21] The 1984 instantiation of Motoman can be found just after Kyle Reese flips on a panel of workshop switches.

[22] Such as some we explore later in this book, including *Battlestar Galactica*.

waddling about on the battlefields of Iraq and Afghanistan since 2007. The problem, though (aside from the fact that he's a robot assassin), is that he still needs some debugging; when he gets stuck or falls over—as he does from time to time—he can't figure out who the bad guy is, and so he just starts shooting in all directions, at all people, and sometimes that makes for what is commonly referred to as *friendly fire*.

Now, according to the movie, the Terminator was first created forty years after 1984. Are we on track to see this Terminator-type technology fully emerge in 2024? I don't think we'll see humanoids, but I think what we'll have will be much more powerful, far less visible, and way more deadly.

One argument for using robots in battle is a moral one: They don't get snagged on touchy-feely philosophical questions. After all, even the best-trained soldier can violate the Geneva Conventions, or the rules of engagement that most armies practice. A 2006 survey conducted by the U.S. Army Surgeon General discovered some surprising trends among soldiers.[23] These trends indicate a mentality and an approach to foreign combat that most Americans might not want to see in action. The survey pointed out that fewer than half of the soldiers serving in Iraq thought that noncombatants should be treated with respect. A fifth of the soldiers claimed that all civilians should be treated as militarily trained insurgents, more than a third believed torture was acceptable, and only half said they would report a colleague for unethical behavior. Ten percent of the U.S. soldiers in Iraq admitted to abusing civilians or unnecessarily damaging civilian property. Tsk-tsk-tsk, engineers reply. Robots, they say, can do better. These problems are what military roboticists refer to as *human error*. Robots are clear in that they're both ethical and reliable. This is why we use robots in battle. Moral reasons.

According to Ronald Arkin, a computer scientist at Georgia Tech, you can remove the human error by removing the humans. Mr. Arkin, who designs software for battlefield robots under contract with the U.S. Army, described in an army report in 2009 some of the benefits of autonomous military robots. He points out that they can be designed with no instinct for self-preservation, which means that they do not freak out when they get scared or injured. He points out that they can be built to show no anger, abuse, or recklessness, so they don't lose their temper. They can be made invulnerable to various psychological problems such as stress, fatigue, religious prejudices, or poor training, so you can't upset them by teasing them. They also don't need food, clothing, holidays, or disability pay.

"It is not my belief that an unmanned system will be able to be perfectly ethical on the battlefield," Dr. Arkin wrote in his report,[24] "but I am

[23] The survey included interviews with 1,320 soldiers and 447 marines in Iraq during the autumn of 2006. Though the report was completed in November of 2006, it was only released on May 5, 2007, after its findings began to leak to the press.

[24] http://www.cc.gatech.edu/ai/robot-lab/online-publications/formalizationv35.pdf.

convinced that they can perform more ethically than human soldiers are capable of." So he sees this as a shade closer to white.

The Army's Computational and Information Sciences Directorate of the Army Research Office understandably agrees, since that department is financing Arkin's work. Randy Zachery, office director, points out that the hope is to "allow autonomous systems to operate within the bounds imposed by the war fighter." And Lieutenant Colonel Martin Downie, a spokesman for the army, noted that whatever emerged from the work "is ultimately in the hands of the commander in chief, and he's obviously answerable to the American people, just like we are."[25]

You can remove the human error by removing the humans.

Regardless of whether it is black, white, gray, autonomous, or controlled, these military specialists are confident that this technology is going in the right direction and that responsible parties will be accountable.

There are other points of view on the military robot debate.

In December of 2008 the U.S. Navy's research division received an extensive report titled *Autonomous Military Robots: Risk, Ethics, and Design.*[26] The authors of the report discuss a rather mysterious 2008 event in Iraq in which several American-made robots armed with machine guns malfunctioned and opened friendly fire on U.S. soldiers. The authors also point to a 2007 incident when an autonomous "robotic cannon deployed by the South African army malfunctioned, killing nine 'friendly' soldiers and wounding fourteen others." The authors outline how hard it is to stop potentially fatal chains of events caused by autonomous military systems, or even systems in cities, homes, and schools that ". . . process information and can act at speeds incomprehensible to us." The authors conclude in their understated style: "It would be naive to think such accidents will not happen again."

What's most interesting about the report is its authors' concluding advice: Give the robots a sense of ethics and morality. The report takes a turn toward the Asimovian by quoting Isaac Asimov's famous Three Laws of Robotics. The authors offer science fiction as a guiding light by which the U.S. Navy can navigate the rather foggy waters of ethics, war, and autonomous robots.

The military is pretty tight with the science fiction community. The U.S. Department of Homeland Security even invited several science fiction authors to invent scenarios and

[25] Both quotes come from the *New York Times* article, "A Soldier, Taking Orders from Its Ethical Judgment Center," By Cornelia Dean, November 24, 2008.

[26] http://ethics.calpoly.edu/ONR_report.pdf.

discuss them at the 2009 Homeland Security Science and Technology Stakeholders Conference, with several of these authors staying on as full-time thinkers and researchers for the department. The forty or so authors, who offered their services in exchange for travel costs to and from the conference, come from a group called Sigma. Through the years the members of Sigma have been invited to attend meetings organized by the Department of Energy, the U.S. Army, the U.S. Air Force, and NATO.

Is The Terminator back? Are we designing ethical and moral decision-making combat robots? When will robots start fighting other robots in real wars? And if autonomy will be given to robots on the battlefield, what decision-making powers will they be given? If foot soldiers are being replaced, are commanders next in line for an "upgrade"? And will Mr. Arkin's autonomous robots eventually replace the generals, too?

Before continuing, I would like to muddy some waters that might otherwise seem clear.

First, a remote-controlled robot is not an autonomous robot. This might appear to be a clear statement, but it's not. The only real difference between autonomous and remote-controlled is at the top level of the decision tree. There's always some autonomy involved in a robot; there has to be in order for it to operate. It's like your car: You don't have to worry about the oil getting transferred to the filter, for example. The car takes care of that autonomously. Your job as the driver of the car (or robot) is to make the higher-level decisions while the low-level stuff is taken care of for you. A remote-controlled robot is just a robot with a very simple (or basic) set of autonomous functions. So the distinction of remote-controlled vs. autonomous is shorthand for determining who makes the big choices. *Autonomous* usually means you can flip it on, stand back, and it'll do the whole job for you. *Remote-controlled* means something more like babysitting. But this varies, and greatly, between different robots.

Remote-controlled robots are currently preferable because they're simpler, easier, and more reliable. These days the U.S. Army has a wide array of these weapons. Two examples are handheld drones designed for disarming roadside bombs[27] and remotely guided missiles developed by Raytheon and Lockheed Martin, which have some limited autonomy for zooming around hillsides, buildings, or other things that happen to get in the way. Maybe the best-known remote control warbot is the SWORDS system, developed by U.S.-based Foster-Miller.[28] SWORDS is essentially a remote-controlled mini tank. It's built like a tank, being both tough and rolling on treads, but you can put weapons,

[27] Created by Honeywell.

[28] SWORDS is an acronym that stands for Special Weapons Observation Reconnaissance Detection System.

grasping devices, cameras, or other stuff on it, and guide it up and down stairs, through bogs and snowdrifts, through sand, etc.

Another notable company working in this field is iRobot, the company that designed and manufactured Roomba, the ever-diligent and ever-cute domestic vacuum robot. Founded by three veterans of the MIT AI Lab (Helen Greiner, Colin Angle, and Rodney Brooks[29]), iRobot is now producing multiple lines of military robots. Some of these robots are remote-controlled, such as the PackBot (another mini tank, like SWORDS), while others are more autonomous systems, such as the X700, an autonomous robot designed for troop relief and urban warfare. "It can be doing re-supply operations, taking ammo or water to troops who are pinned down, [or run] perimeter security and building clearing," Greiner, the company's chairman and cofounder, told the *Army Times* in October of 2007. The X700 can be weaponized with an electronic firing system that is able to shoot as many as sixteen rounds a second.

"His weapon is not at his shoulder; it's up to half a mile away."

Someone using these systems can be sitting a kilometer away, rigged up to a control console where he's able to look through a camera and see what the robot's camera sees. Each of these systems sends data back to the soldiers at the other end of the control unit, so it's clear that the goal has been to build a remote weapon, not an autonomous one. According to Bob Quinn, a manager with Foster-Miller, the only difference to a soldier is that "his weapon is not at his shoulder; it's up to half a mile away."

There are long assembly lines rolling these robots out as you read this sentence. Military robots, like machine guns, are becoming another part of the automatic weaponry industry. The Cormorant is a nautical warbot that submerges to a 50m depth for monitoring and attack. The PANDA[30] is used to track down Somali militants. The SGR-A1[31] uses a camera and image-recognition technologies to pinpoint targets up to 500 meters away. Israel has armed robotic shoeboxes along the wall of the Gaza border. The Wasp is a radio-controlled flying "insect" weighing a mere 170 grams, used for surveillance. And there is also the MATILDA,[32] another tank-like rolling system that is so close to SWORDS that I continually get both of them confused with the PackBot.[33]

Even though these are remote-controlled systems, there is still some level of autonomy,

---

[29] Rodney Brooks is also former director of the MIT Computer Science and Artificial Intelligence Lab. Before he joined the MIT faculty, he held research positions at Carnegie Mellon University and MIT, and a faculty position at Stanford University.

[30] PANDA is an acronym for Predictive Analysis for Naval Deployment Activities.

[31] Made by Samsung Techwin.

[32] Made by Mesa Electronics.

[33] Made by iRobot.

and things can still go wrong. Some of these military robots have misfired, killing civilians, students, training soldiers, and pedestrians.

Let's go back to South Africa and the misfire mentioned above. This is probably the most famous and tragic instance of an automatic system swinging guns in the wrong direction. In October of 2007, The Oerlikon GDF-005, an automatic and semiautonomous antiaircraft gun, managed to get out of the remote control of its users and began spraying over 500 rounds of ammunition, including hundreds of high-explosive 35mm cannon shells in several very bad directions. The robot eventually left nine soldiers dead and wounded fourteen. Media reports[34] stated that the war-games exercise, which used live ammunition, took place at the SA Army's Combat Training Centre, at Lohatlha. South African National Defence Force spokesman Brigadier General Kwena Mangope told South Africa's *The Star* that it "is assumed that there was a mechanical problem, which led to the accident. The gun, which was fully loaded, did not fire as it normally should have," he said. "It appears as though the gun, which is computerized, jammed before there was some sort of explosion, and then it opened fire uncontrollably, killing and injuring the soldiers."

In similar events, multiple M-series chain guns (mounted turrets that are capable of rapidly firing .50 caliber and 3mm explosive ammunition) have malfunctioned due to any of a number of problems, ranging from locked servos, looping kill-switch safety algorithms, satellite feed lag, to that long-enduring favorite, human error.

Human agency is gradually being removed from the equation of war.

These stories present ethical tangles to be combed out in military robots. For example, when a military robot malfunctions, who's responsible? What if the robot was totally autonomous; who's responsible then? Or, if it's remote, is the robot controller the guy in control? Do we blame the mechanical engineers? Or is the occasional algorithmic backfire the programmer's fault? Is it the manufacturers—are they responsible because they made the thing? What about Congress? If Congress allowed these things to be commissioned, contracted, constructed, and deployed, then aren't they the ones responsible? Or is it the president? Or is the robot the one to blame? And if it's the latter, then should a robot be allowed to go against orders? Should it run on Windows, Linux, or Mac OS X? [35]

Autonomous systems are a means of delegating decisions, and thus, responsibility. As this

[34] ITWeb, a South African technology and IT news source, Wired Danger Room, *Mail & Guardian*, and Gizmodo.

[35] No, seriously; those are the operating systems that are used for many of these robots.

happens, human agency is gradually removed from the equation. With it goes our sense of moral obligation. This means that it is easier to instruct pilots to kill, simpler to perform, costs less, takes less time, and requires almost no risk. Like putting down dogs in a pound, it's more humane.

There are voices of reason in this wilderness; they at least help us to hear the problems. Patrick Lin, one of the authors of the 2009 U.S. Navy–funded report, brings up some of these questions, and writes, "There is a common misconception that robots will do only what we have programmed them to do. Unfortunately, such a belief is sorely outdated, harking back to a time when . . . programs could be written and understood by a single person."

BigDog, a military robot.
Courtesy of Boston Dynamics © 2009

Another of the voices against autonomous military systems is Professor Noel Sharkey's, of Sheffield University. Sharkey proposes a global ban on autonomous weapons until they can comply with international rules of war, especially the one that prohibits the use of force against noncombatants. This is very close to a real solution to many of these problems, provided countries even recognize that robotic warfare exists; that they understand the long-term consequences of warbots; and that they actually follow the prohibitions—and, provided they can agree on the definitions of *robot, remote-controlled*, and *autonomous*.

In the United States, as in most developed countries around the world, the military robotics industry is growing. The support of the U.S. government causes entrepreneurs to flock to the financial fat barrel.

Unmanned and autonomous military systems are on the rise. In 2001, Congress mandated that a third of all military ground vehicles must be unmanned by 2015. According to the "Unmanned Systems Roadmap 2007–2032," the Pentagon planned on spending over $4 billion by 2010 on unmanned systems, also increasing the autonomy of these systems so that troops spend more time monitoring systems than anything else. In July of 2009, the U.S. Air Force published its "Unmanned Aircraft Systems Flight Plan 2009–2047," predicting the deployment of fully autonomous attack planes. The report states: "Advances in AI will enable systems to make combat decisions and act within legal and policy constraints without necessarily requiring human input." According

Courtesy of Boston Dynamics © 2009

BigDog, on his way up.

to *BusinessWeek*,[36] the U.S. military plans to triple its inventory of unmanned aerial vehicles (UAVs) in the next ten years. In 2010 the Pentagon will produce more UAVs than manned, and train more UAV pilots than traditional fighters and bomber pilots combined. As General David Petraeus, head of U.S. Central Command, puts it, "We can't get enough drones."

The reason for this growth in military robotics is financial. Wars usually end because humans have died. The country isn't motivated to stop a war if none of its citizens are getting killed. As the war in Iraq progressed, American media focused with the most concern and precision on the number of American soldiers that died, not the number of Iraqis. When American casualties reached somewhere over 3,000, calls to stop the war started gaining a sympathetic audience. Same scenario with American casualties in Vietnam. So if sending in a small army of warbots to a foreign country just means that a few robots get blown up, there is hardly a motivation for wealthy and powerful countries to curb their inclination to invade from time to time. It just becomes good television.

In fact, robots getting scattered on some battlefield actually becomes motivation to continue a war. The more robots someone blows

[36] "Pentagon to Increase Stock of High-Altitude Drones," by Tony Capaccio, February 5, 2010.

up, the more the military gets to build, and the more they get to build, the deeper into the fat barrel they get to scoop. So if our robots are getting blown up in a war overseas, we've got a pretty good reason to continue making more robots. Not only is robotic war cheaper, easier, more effective, and less emotionally taxing, but scattered robot parts also help goose innovation. It makes the guys shooting at the robots happy, and it makes the guys building the robots happier still.

Just as with the American automotive industry in the 1950s, this roboticization of labor may have other impacts. Recruitment centers will be affected, and the young GIs that sign up for a new career will be replaced by robots in the field. Those soldiers that might have otherwise collected a salary will have to find other roles in the national economy that don't involve shooting or getting shot at.

Warbots, like machine guns and armored vehicles, are automatic weapons. They are a significant step in the evolution of warfare. According to Peter Singer, author of *Wired for War: The Robotics Revolution and Conflict in the 21st Century*,[37] more than forty countries are developing UAVs or other military robotic systems. They include India, Pakistan, Belarus, Iran, Georgia, and Russia. China, of course, is on the warpath, and has already rolled out the Pterodactyl and Sour Dragon (respective rivals to America's Predator and Global Hawk). The

Courtesy of Boston Dynamics © 2009

BigDog gets muddy, too.

Israelis also have something called Harpy—a fully autonomous UAV that dive-bombs radar systems with no human intervention. I've even heard weird stories of Koreans trying to make fembots that will be used as psychological weapons against the Chinese.

Military robotics weapons bans will be very difficult (if not impossible) to enforce, because the technology is inexpensive, flexible, easy, and portable. You, too, can suck from the deep fat of the pork barrel named military robotics. With a local Radio Shack, open-source software, about $500, and a free weekend, you, too, can build your own battlebot. Public Web pages document do-it-yourself autonomous weapons systems.[38] One is built with motion detection

[37] Peter Singer is also director of Brookings's 21st Century Defense Initiative.

[38] http://www.diydrones.com/ is a good one for UAVs; another is at http://members.upc.nl/a.kutsenko/guide.htm.

BigDog can also be helpful for your backwoods skiing trips.

software that positions moving objects in the camera's frustum. That software is attached to a servomotor controller. All of it is available for download and implementation with parts from Radio Shack (except for the gun). As I understand it,[39] this robot is as capable of shooting a .40 caliber Glock as it is a Namco water pistol.

Robots Don't Kill People; People with Remote Controls Kill People

ROBOTS PRESENT MORAL PROBLEMS. BECAUSE A robot (military or otherwise) can affect how another person behaves, there are ethical, moral, and legal boundaries that need to be clarified.

Even if military strategies are informed by science fiction authors or AI specialists, human decisions drive the technology, not vice versa. Despite how it would like to portray itself, an army is not a machine; it's a social structure. So humans, politics, and ambition have more influence than technology.

The militaries of most countries—including the U.S., Korea, China, Australia, Canada, and more than a dozen others—simply can't follow the rules of engagement, which is one reason why Asimov's Three Laws can't be implemented. The argument each country makes is basically the same: "The rules of engagement are

[39] I've neither seen nor operated it, myself.

obsolete because {*enter enemy's name here*} won't follow them either, and anyway, we might get attacked first." Yes, it is true; some kids in Amsterdam might go down to Radio Shack with €500 and start making their own robots. And to be quite frank, it would cost much less than €500 to make a lethal robot.

So much for Asimov's laws, and so much for utterly unenforceable UN charters.

Might will always make right. Who controls it is the question.

What's to be done?

Briefly, there are two solutions to this problem.[40]

First is open-sourcing a robot operating system. This means that everyone owns it, everyone knows how it works, and everyone can operate it. This also means that ethical and moral problems will become clearer more quickly. This solution is a bit like the U.S. Constitution's Second Amendment. There were multiple motivations behind open-sourcing firearms, but the amendment reduced the possibility of a foreign invasion and ensured that the majority would stay in power. The motivation I like the most is making sure that the government is kept in check. An open-source robotic operating system is the Second Amendment's knowledge-based cousin, which minimizes centralized power, thereby reducing the likelihood of war. It's different from the Second Amendment in that it distributes knowledge first, and weaponry second.[41] If the most-advanced technologies are a public resource, then the means, methods, and threats that the technology presents could be countered and controlled in the checks and balances that democracy will hopefully continue to offer. This would put the technology in everyone's hands, and it would at least be something that would keep these machines, and their remote controllers, in check.

An argument against this solution: "Oh, no. That would cause a weird arms race—first, between governments and citizens of industrialized nations, and second, as the technology spread, between nations in general." To put it as simply as I imagine it to be, that was precisely the intention of the Second Amendment. Might will always make right. Who controls it is the question.

A universal operating system for robots will most likely evolve just as personal computers did. There is the Apple way, in which a quality OS is built for proprietary hardware, or there is the Microsoft way, in which a mediocre OS is built for ubiquitous hardware. Then there are the open-source systems in which, like the Internet, collaboration creates a shared technology.

[40] I propose these in shorthand, and I still think Sharkey has some good ideas, too. Time will tell, ultimately.

[41] Note that knowledge about how a warbot works is at least a few steps away from implementing one; that's important to this strategy, and quite different from the Second Amendment.

Linux and other open-source prospects seem a third solution that has some distance to travel if it hopes to handle all the different kinds of hardware types that robots present (rolling, walking, gripping, etc.), but it is probably the best bet for a widely used robotic OS. Although there is currently not enough market demand, now is a good time to start thinking about it.

This approach is being explored by companies like iRobot—probably the single most successful robotics company to date—which has published their robot operating software package as open-source. Gostai, of France, and Willow Garage, of the U.S.—two other companies that produce robot operating systems—have also published their software as open-source. Various editions of the GPL[42] free software, or open-source, license allow companies to both produce, profit from, and share their software with development groups and the general public. What these companies and research labs lack, however, is a binding force to help them orchestrate, distribute, and manage the production of (not to mention debug) these robot operating systems.

The second solution to the escalation of warbots is to put checks and balances on the military that uses them. Installing checks and balances seems to be the approach of the American Civil Liberties Union (ACLU), which, in March of 2010, filed a Freedom of Information Act lawsuit against the U.S. Defense Department. The lawsuit demanded that the government disclose the legal basis for its use of Predator drones to conduct targeted killings. This lawsuit was especially sensitive since targets not only included Afghanis, Iraqis, and Yemenis, but American citizens, as well.[43] The lawsuit specifically asked for information on when, where, and against whom drone strikes can be authorized, arguing that CIA agents should not be piloting drones to attack terrorism targets, as that approach clearly does not fall within the internationally agreed-upon rules of war.

A new arms race is heating up. While it will, like the first cold war, be a race for technological dominance, it will not be a silent amassing of technology centralized within two competing governments. It will be distributed globally; it will emerge in many surprising and often invisible ways; and it may be as significant and as world-changing as the invention of gunpowder. As enemies become more invisible, surveillance becomes more important. In an inversion of usual tactics, robotic defense will be met with robotic offense. A man sitting in his living room in the Middle East will suddenly be covered by small metal dots, and when the dots quickly disappear, he

[42] GPL, or the GNU General Public License, or just GNU Public License, is a free software license, originally written by Richard Stallman for the GNU project.

[43] http://www.aclu.org/national-security/aclu-v-doj-et-al-complaint, http://www.aclu.org/national-security/aclu-seeks-information-predator-drone-program.

will fall dead to the floor. Drones will not only send bombs down, but they will also be used to bring data up, as well as packages, paperwork, and abductees. Drones are becoming smaller, changing from airplanes into hummingbirds. Tiny robots are embedding themselves in mountains and fields, forming grids in the Earth, grids in the air, and grids in the walls. Even smaller robots are making appearances as houseflies.[44] Flexible blobs are squeezing under doors.[45] Particles are hovering near lightbulbs. Any surface may be a surveillance device, any object a bomb.

Nothing improves the killing process more than automation.

Whether they are in the front yard of a house or the foothills of a war theater, robots are effective killing machines. Nothing improves the killing process more than automation, especially remote automation. Ask any of the great military inventors or despots of the last hundred years, be it Gatling or Hitler, and they'd tell you that automating a process cuts costs, ups productivity, and generally improves efficiency. In *The Terminator,* Kyle Reese and Sarah Connor are in a garage, talking about Reese's childhood and the prevalence of machines built in automated factories. Reese says, "Most of us were rounded up, put into camps for orderly disposal." He then rolls up his sleeve and shows Sarah an identification mark on his forearm that looks quite a bit like a Nazi identification number from World War II. "This is burned in by laser scan. Some of us were kept alive, to work, loading bodies. The disposal units ran night and day. We were that close to going out forever."

Well, that's enough of that. If you'd like to read more, I'd recommend *Wired for War*.[46]

A Quick Stroll through the Uncanny Valley

TO COME BACK TO PRESENT-DAY REALITY FOR A minute, you won't be mistaken for bacon and you won't be shot at, bombed, or trapped because you triggered a robot's sensor as you walked across some foreign field of war. Perhaps the chances are a little better that your neighbor's lawn guardian home-brewed robot will get you, but not much. Warbots won't be a threat for most of us, and we just have to wait and see how humane we will be as we shoulder our new robotic avatars of battle.

But there *is* something for you to legitimately fear.

As robots explode into existence around us, evolving into oil rigs underfoot and cloud seeds overhead, into nanotech assemblers that creep up between the cracks, and galactic explosions that rain down from above, we

---

[44] See chapter 8 for more on this.

[45] Part of a $3.3 million project funded by DARPA, iRobot's blob robot expands and contracts to move under doors or other tight spots.

[46] For much more on this, see Peter Singer's book, *Wired for War: The Robotics Revolution and Conflict in the 21st Century.*

mere mortals that still have to shit, shower, and shave in the mornings have become, simply, more ignorant of what we're making. Our fear is based on ignorance. We just can't keep up anymore. Even our present is a mystery. We can no longer keep track of our technologies. Our cultural evolution has surpassed our physical evolution, and that's spooky, because as technologies advance we become more ignorant every day. And since fear is always based partly on ignorance, we are more fearful of our technologies than ever before. As the great Douglas Adams put it,

> *Anything that is in the world when you're born is normal and ordinary and is just a natural part of the way the world works. Anything that's invented between when you're fifteen and thirty-five is new and exciting and revolutionary and you can probably get a career in it. Anything invented after you're thirty-five is against the natural order of things.*

Apart from ignorance, we're afraid of robots—and androids in particular, because they look like humans. One of the great horror themes in science fiction is large androids walking down the streets of Los Angeles, stalking poor, innocent Sarah Connor, their red eyes blazing under a hood of hair and skin, toting guns, and coming on like unstoppable rapists from the future. They blend in with us, act like us, and are able to do all we can do, only better. Humanlike robots are frightening because, as the record shows, humans are pretty dangerous animals.

I can't think of any other species walking around on the planet today—or any day, for that matter, that's more unpredictable, insidious, or deadly. We're first-class predators. This is why we have our eyes on the front of our heads—for depth perception and better telescopic targeting, same as hawks and cats. Every time you look a human in the face you're looking at a predator. We're so far up the predatory food chain, in fact, that we're now a threat to ourselves. So if you ever want to really scare someone, put a human face on an unknown technology and roll it out into the middle of the living room while they watch television or read a book. Robots are a mirror, and they give technology a predatory face.

Our fear is more than just face-deep. It goes back to well before we had steel and wheels. Our fear of robots—and specifically androids—is triggered by an instinct of wanting to protect the tribe. It's a social survival response. "Will this creature threaten my family?" "Will it take over my city?" Or, "Will it eat my wife because it thinks she is bacon?"

This instinctual fear is called "the Uncanny Valley," a famous theory proposed by roboticist Masahiro Mori in 1970.[47] It says that when we're around a robot that's almost human-looking (but not exactly), we get spooked. If the robot

[47] Which Mori outlines in *The Buddha in the Robot*, among other writings.

The Uncanny Valley

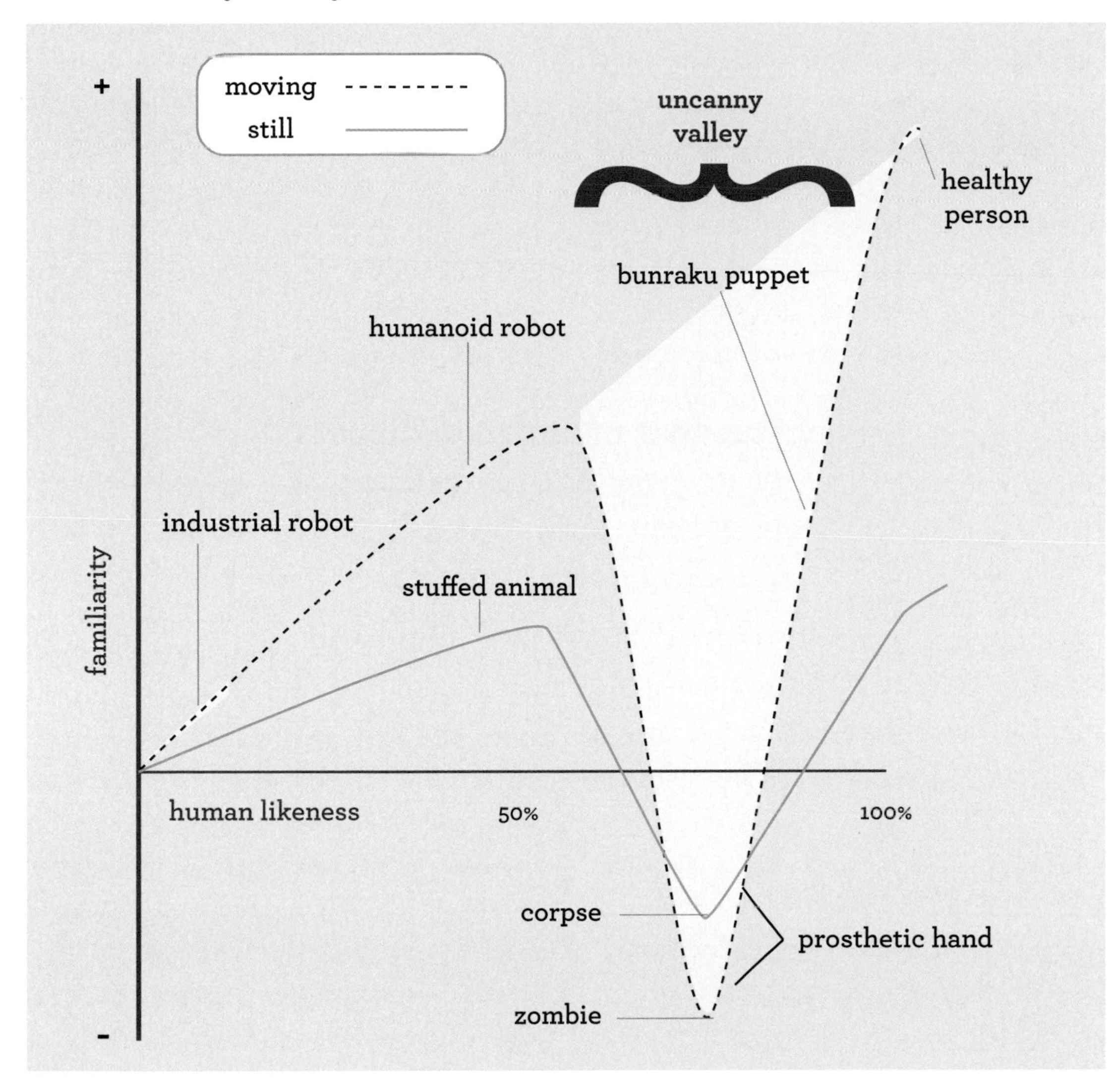

doesn't look anything like a human, that's fine, but when it starts to look like a human, that's when we get freaked out.

Mori says that fear of humanlike robots is generated by two factors: appearance and movement. If a robot's humanlike face is even the slightest bit strange, we would notice. If the skin is a little too waxy, if the face is slightly disproportionate, if the eyes are matte instead of gloss, if the wig is a little bit discolored, if the lips are the wrong color—these things would be noticed and would trigger, on an instinctual level, an odd feeling. Not only in the face, but the body, too.

The second robotic fear factor is movement, which, like other visual factors, can both trigger or add to the existing uncanniness. An example of this is seeing someone with a twitch, or someone that walks with a slight limp, just subtle enough to dwell at the periphery of your attention.

Karl F. MacDorman is associate professor in the Human-Computer Interaction program of the School of Informatics, at Indiana University. He is also something of an Uncanny Valley tour guide. He and several colleagues documented the emotional responses of 140 test subjects who were asked to view various androids. The results indicated that androids from the Uncanny Valley were linked to fear, shock, disgust, and nervousness. MacDorman suggests that the Uncanny Valley response is linked to a "fear of one's own mortality [and an] evolved mechanism for avoiding pathogens."

Our fear of androids has to do with not just a fear of getting sick, but a fear of losing bodily control—a fear of death itself.

I conducted an e-mail conversation with Dr. MacDorman for several months, and he set out a broad range of causes for this response. He explained that we all instinctively try to pick a proper mate. We are all wired to be attracted to healthy individuals for simple reproductive reasons. The flipside of this means that if we find someone that is not healthy, we avoid them like the plague. MacDorman calls this a *pathogen avoidance response*. But wait, there's more. He says that the Uncanny Valley also involves a kind of empathy, where if the robot is almost human, but not quite, it trips a neurological wire and creates a kind of mental dissonance, a bit like you might experience if you look at a chimp doing something very human (only the opposite, really). MacDorman's tests also point out that when we see an android and get that uncanny feeling, the robot violates beliefs of normalcy, both cognitively and culturally. Our fear of androids has to do with not just a fear of getting sick, but a fear of losing bodily control—a fear of death itself. And lastly, religion is involved, including our spiritual ideas about what it means to be human, creating nothing less than a tiny existential firecracker in the head of the viewer.

All this from looking at a robot. No wonder James Cameron made a horror movie about an android-cyborg assassin.

It's complicated and instinctual. Princeton University researchers found evidence that even macaque monkeys may experience the Uncanny Valley.[48] They found that when looking at

[48] Specifically, Asif Ghazanfar (an assistant professor of psychology at the Princeton Neuroscience Institute) and Shawn Steckenfinger (a research specialist in Princeton's Department of Psychology). See the October 2009 edition of the *Proceedings of the National Academy of Sciences*.

Mark Stephen Meadows

A little dog! Not recommended for combat missions. (Sony's Aibo, designed by Sorayama.)

Mark Stephen Meadows

One of Sony's Aibo designs. National Museum of Emerging Science and Innovation.

computer-generated images of monkeys that look similar to them, but just a little bit off, the monkeys' reaction was strangely similar to that of humans. Normally the monkeys coo and make kissy sounds when they see one another, a bit like a greeting. But when the monkeys were given images of other monkeys that were just slightly distorted, they would first avert their eyes and act a bit scared, and then, when given the chance, stare at the images for longer than normal. It is also well known among Aibo owners that dogs do not care for the Aibo, either, so perhaps dogs also experience the Uncanny Valley.

The Uncanny Valley is certainly open for human habitation, too. There have been more than a few moments in Los Angeles when I was sitting at a table, eating my own lunch and minding my own business, when a woman, usually wealthy and white, would walk in and sit down at some nearby table. I would suddenly pause, fork suspended in front of my mouth, as I noticed a lack of muscle around the lips, a flat and waxy skin tone, something uncanny, and that tiny existential firecracker would go off in my head—all because this poor woman was the victim of a plastic surgery misdemeanor. As we humans modify ourselves and adopt new technologies directly into our bodies, we will slip over the hill of canny into a new valley of uncanny, only explored by the unfortunate victims of scalpels, silicone, car accidents, and vanity. This is a precursor to humans becoming more robotic—not a robot bordering on human, but a human bordering on robot.

The Uncanny Valley is scary stuff.

Visiting with robots today and watching these horror shows reminds me of when I was a kid. My friends and I would have slumber parties in an old cave. We'd sit in there with a flashlight and shine it on the wall, telling ghost stories. The flashlight was the most important part of our show. It was used to tell the story, always, with hand gestures and shadow puppets on the wall, and then, at the final moment of the story, when we were all pretty scared and ready for the big conclusion, the storyteller would shine the flashlight on his face, and we'd all scream like little girls as we ran out of the cave.

These things aren't dangerous, but they sure are scary.

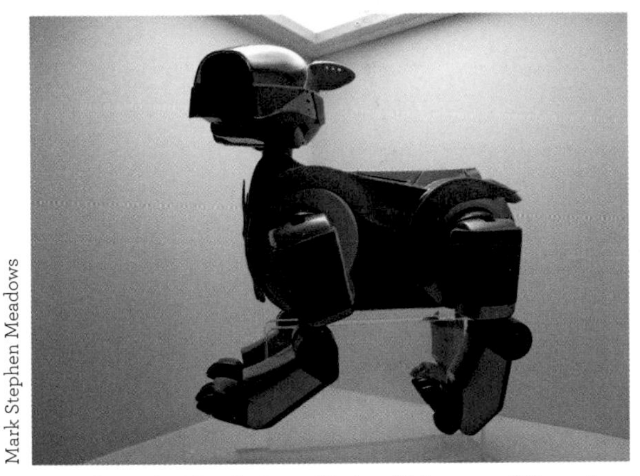
Mark Stephen Meadows

One of Sony's Aibo designs. National Museum of Emerging Science and Innovation.

Why I'm Afraid of My Own Shadow

OUR FEAR OF ROBOTS IS A SOUP WITH THREE main ingredients: fear of humans, fear of technological advancement, and an instinctual self-preservation mechanism that can get rubbed the wrong way.

Though the database is mightier than the gun, military robotics is still a dangerous technology. While robots with faces are strange and cause fear, they're not dangerous. At least, not yet. So the military today represents a danger, and a fear, that we need to have of robots. But the fear is not one of being shot; instead, it is one of backfiring decisions, poorly aimed intentions, and explosive politics. The robots themselves are not the problem—humans are, and the information that robots provide is a danger that is even more important than military concerns.

We humans are the problem because we humans run the robots.

Fortunately, we humans are also the solution. As James Cameron remarked:

> *I don't think anything resembling the Terminator is really going to happen. [*Terminator *is] about us fighting our own tendency toward dehumanization. When a cop has no compassion, when a shrink has no empathy, they've become machines in human form. Technology is changing the whole fabric of social interaction.*[49]

[49] http://www.wired.com/entertainment/hollywood/magazine/17-04/ff_cameron.

Chapter Two: The Jetsons

On Robots That Cook Pancakes, Vacuum, and Tend to Cities, Farms, and Fields—Interaction, Part 1

> To you a robot is a robot. Gears and metal; electricity and positrons. —Mind and iron! Human-made, if necessary human-destroyed! But you haven't worked with them, so you don't know them. They're a cleaner, better breed than we are.
>
> —character Dr. Susan Calvin, Robopsychologist, U.S. Robots and Mechanical Men, Inc., in *I, Robot* by Isaac Asimov

IT WAS A SUNNY SUNDAY MORNING AND I WAS doing the best I could to clear my head from the jet lag. Between the coffee, painkillers, and fried chicken wads I had eaten, I was feeling like a champ. No, really, it was a great combination, and left me wanting nothing. But the jet lag had spun me a bit.

I shut off the computer and opened the curtains to look outside. I was treated to the panorama of a row of windows that ran up a building that was exactly like the one I was looking out from. I saw a man directly across from me sitting at a desk, typing. It looked cold outside. I closed the curtain.

I pulled on some leathers, rinsed out my mouth, put on some earphones so I could listen to my music, took the elevator down twelve floors, walked through the lobby, smiled to the nice man at the hotel desk, and stepped into the Tokyo street.

I was going out to visit *Japan!* It was like visiting the cultural equivalent of a giant magnet. What other city in the world could be weirder? Where else would an international condensation of ideas mentally assault me? Where else could I find the limits of reality?

Physically Active, But Shy

GETTING TO A NEW CITY AT NIGHT DOUBLES YOUR fun. You get the experience of arriving once, in one world, and then you get to see it again the next morning with entirely new eyes. If nighttime Tokyo has the flavor of *Blade Runner*, daytime Tokyo is *The Taste of Tea*. There's something so quaint about the city, something so friendly and clean, so cute and colorful, that I couldn't help but smile, turn up my music, and walk a little more smoothly among kids on bikes, an apple-faced little old woman whose eyes sparkled with all the intensity of eight

Mark Stephen Meadows

Tokyo is a world that has it all: the past and the future, tradition and modernity, funk and family.

decades, tiny people that carried huge sacks, and a family that looked so healthy and happy I felt like I was looking at a supermarket advertisement. It was an ideal world. Between the edge of night and the break of day, machinery and family, Tokyo is a world that has it all: the past and the future, tradition and modernity, funk and family. It is somehow perfect on the inside. Then again, maybe that was the codeine. Either way, I loved it all. The walking was doing me some good. There was even a little park with a big red arch that indicated an outdoor temple. I found some statues of Buddha, and incense burning. Big crows hung out around there and watched my every move, coughing out strange words in yet another ancient language that I couldn't understand. The park was intensely peaceful and felt quite old. It helped me get a sense of what was going on.

There were no robots, no machines, no industry, automation, or media, and I was enjoying the break. But after an hour among the talkative old birds and silent statues, I was getting cold. I considered getting more chicken nuggets (or whatever they were), but as I exited the park I realized two things: One, I was totally lost.

And two, I had forgotten to bring my map.

On the night of September 23, 2008, a unicycling robot named Seiko-chan[1] was born under the astrological sign of Virgo. The little[2] robot—which Murata Electronics says is modeled after a female kindergarten student—has a pair of gyroscopic sensors that can measure the angle at which she's rolling and help her balance.

She rolls like Rosey the Robot, from *The Jetsons*.

The single wheel moves the robot forward, back, and with a slight angle and a twist to the side, the robot is able to efficiently spin and move forward. I'd always thought that this was some science fiction conceit, something weird the writers of *The Jetsons* had spun up that wasn't based in reality, but in fact, it's a very efficient means of locomotion. The rotating flywheels in the torso help turn the unicycle and maintain balance; all that needs to be done then is to propel the robot forward and back. Seiko-chan is able to move around obstacles by using ultrasonic sensors that measure distances and calculate her speed. Meanwhile, Seiko-chan has an embedded camera that transmits live video via Bluetooth. According to the press release, Seiko-chan is "physically active but shy," and her dream is to "[travel] around the world with MURATA BOY."[3]

This marketing campaign worked beautifully. Both robots are very famous in Tokyo, and in the metro stations you can find enormous wall-sized ads that show MURATA BOY pedaling his bicycle near a wall. Around the corner from

1 http://www.murata.com/corporate/boy_girl/girl/index.html.
2 50 centimeters (20 inches) tall, 5 kilograms (11 pounds).
3 MURATA BOY was a small bicycle-riding robot that was introduced the year before, also by Murata.

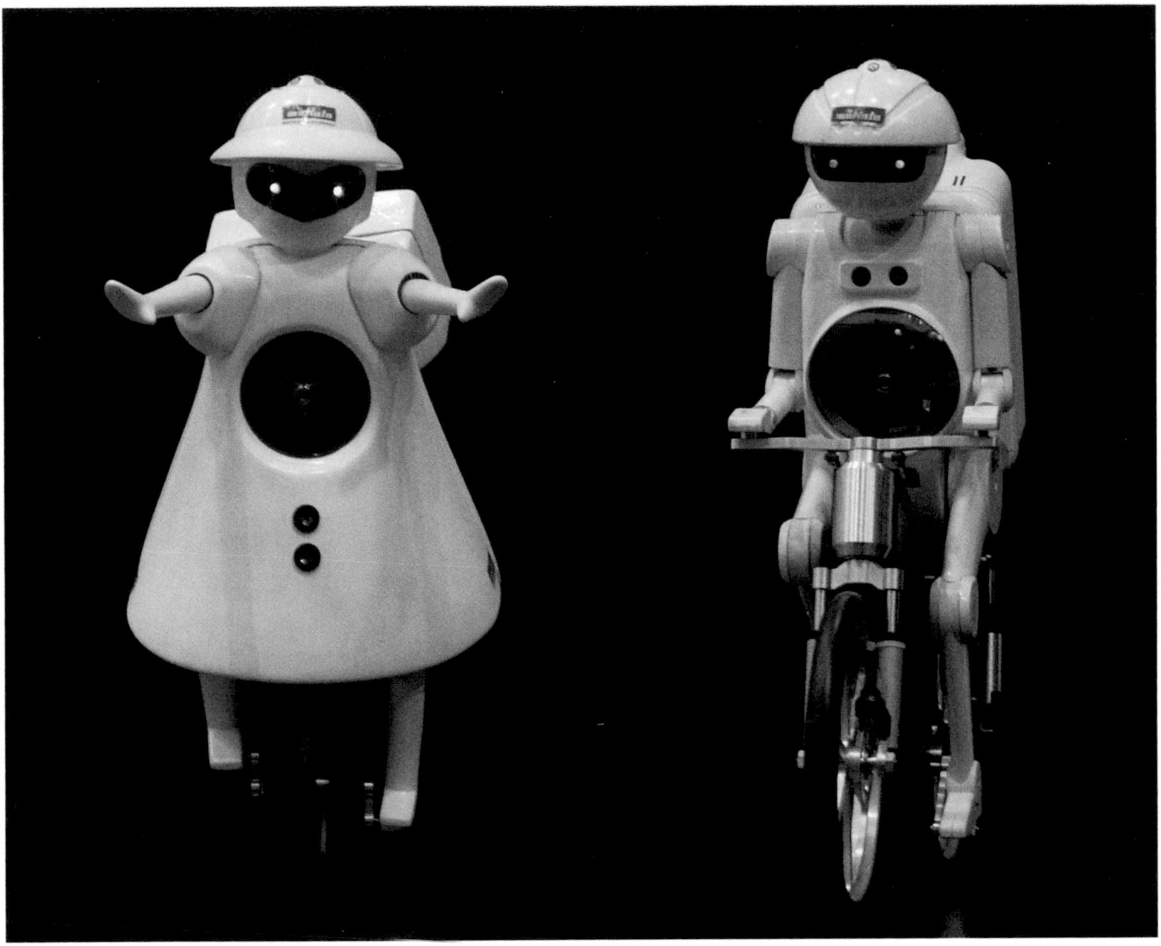

Pink Tentacle (http://pinktentacle.com)

Seiko-chan and MURATA BOY.

him is a young woman, dressed in a futuristic-looking outfit, who seems engaged in a kind of hide-and-seek. The robots are also seen frequently on television.

It was about one o'clock in the afternoon, and I was still lost.

The large buildings had been hemming me into small alleys of traditional Japanese huts, little neighborhoods somehow tucked in among the skyscrapers. I stepped out into a broad shopping center, and this was when I came across a cluster of about a hundred people. I'm rarely attracted to mobs, but lacking direction or reason, I turned my music a little lower and strolled over to see what they were looking at.

They were all watching this robot, Seiko-chan.

I flipped off my music in a delirium of glee. My first robot, and in the wilds of Tokyo! I was so happy. In that moment of glee I barely noticed that I was in front of the Shinjuku metro station, a few blocks away from my hotel. I was enraptured by the appearance of this robot. It was like searching for a critically endangered species in the forest and finally discovering it, after thousands of miles on airplanes and trains. A real robot!

I reached for my camera, then realized I'd left it in my hotel room. As if teasing me, Seiko-chan rolled up to me, stopped, spun a circle on her one wheel, seemed to look at me, and then rolled quickly away in a spurt of dust and a quiet whir of servos.

The Robot Maid, Part 1

EXACTLY FORTY-SIX YEARS (TO THE DAY) BEFORE Seiko-chan was born, something strange entered the American household. It was on the night of September 23, 1962.

ABC was broadcasting a glossy new high-tech, prime-time, made-for-the-whole-family, ultramodern television show: *The Jetsons*. It was ABC's first color broadcast series. This was a rather revolutionary step, a bit like high-definition or 3-D television today, but probably a bigger leap at the time.

The producers at ABC knew that this show would attract viewers who were up-to-date on new technologies, drawing in the science fiction fans and other propeller-heads of the day. After all, it was a show that took place one hundred years in the future, so it was a bit of a "new media" show for the forward-looking tech geeks of the 1960s. Even cooler, the show featured a robot maid. Robots hadn't been broadcast via color television before. So the show became not just a simple television broadcast, but a show of the future presented with the most high-tech media available at the time.

The Jetsons was about a relatively standard family's relatively standard life, in the year 2062. There was a father named George, a mother named Jane, two kids, a dog, and that robot maid. It might have been *Leave It to Beaver*, or *Hazel*, or some other show that fit the black-and-white format of the day—some other insider's view into the tiny dramas of a typical family.

The show was not destined to last for long in prime time because, at the same time, on a different channel, Disney was broadcasting competing high-color content. Unfortunately for *The Jetsons*, most viewers chose Disney. *The Jetsons* got moved to Saturday mornings, beginning a trend that saw cartoons relegated to kids' viewing schedules. Nevertheless, this little science fiction foray, despite withdrawing into the puerile corner of Saturday mornings, left an idea in the heads of Americans, and that was the idea of the domestic robot.

Meet the Jetsons.

Cuckoo's Nest/Hanna-Barbera/Wang Films / The Kobal Collection

The series is about a fairly normal family, the Jetsons, as they dance through their mundane family affairs. Three days a week, a three-hour workday amounts to pushing different buttons on different machines (until their fingers get bent from overuse, a precursor to today's carpal tunnel syndrome), and then after the workday, a treadmill is used for daily constitutionals. School kids go for field trips to Siberia. George Jetson commutes in a flying space car that collapses into a briefcase. Meals are prepared by pushing a button. Other families live near the Jetsons in other Seattle Space Needle–style apartment buildings. They do the same everyday things as the Jetsons. The world is automated and hygienic, and the family's technology amounts to a super-clean modernity that only a plastic world, inspired in the early 1960s, could imagine.

And families have domestic robots.

The Jetsons have one named Rosey; in fact, the series began with her. In the first episode of the first season, Jane Jetson wants a maid. She's tired of pushing buttons (washing, ironing, and vacuuming each require pushing a different button), so she wants a robot to make her life simpler. That afternoon Jane heads to "U-Rent-A-Maid" to take advantage of a free one-day trial deal. U-Rent-A-Maid, a shop where one buys used robots, is a rather sleazy operation. The used-robot salesman is a 2062 version of a used-car salesman. The first model he shows Jane is "an economy model from Great Britain" named Agnes.

Developments are not far afield from what the writers of *The Jetsons* predicted.

Agnes is dressed in the classic English attire, repeats "hip hip" with all the cheeriness a robot can possibly muster, and carries a platter of tea and something that looks like mince pies. She even has, under the chipper ring of her metallic voice, a distinct BBC accent. She's not the right model for Jane, though; too rigid, too classic—too snotty, perhaps.

Jane asks to see another model. Next, the used-robot salesman offers a "fun, runabout model that's imported from France," named Blanche Card. She rolls in accompanied by wild sexy saxophone music. She's a curvy, cute fembot with big eyes, big lips, and what can best be described as Pamela Anderson's figure stuffed into a French maid's outfit. The salesman points out that she's "chic" (Jane replies, "Very cheeky") before going on to remark that "The engine is in the rear—where the engine belongs!" Blanche, sex slave that she might be, pivots and shows us her can. She's not right for Jane, either.

"You mean, that's all?" Jane asks.

"Well, we do have one old demonstrator model, with a lot of mileage," the salesman says. He turns his head and yells, "Rosey!"

Offscreen, a motor putts to life, and Rosey rolls in, wearing an apron. At forty-five years of age she is well outdated.[4] Unlike the other two models, she does not repeat herself, and she seems neither British authoritarian nor Pamela Anderson sexpot. Jane tells the salesman she'll take Rosey, who immediately climbs the wall with joy, reaches the ceiling, and falls on her head with a crash. The salesman says, "We can't guarantee her, you know." Rosey responds by sticking out her tongue (a flat, reddish thing that looks like a tongue depressor from a doctor's office).

[4] Model #XB-500.

Rosey quickly becomes a member of the family. She performs her duties around the house, like cleaning and doing laundry, and she appears to be a damn fine cook, too. Her emotional involvement in the family proves to be combustible, perhaps more than any other member. In the twenty-sixth episode (Season 2), she runs away because she thinks (mistakenly) that the Jetsons are going to replace her. Rosey disciplines both Elroy and Judy at times, and she gives advice to George and Jane on such subjects as household management, raising kids, and marriage. Rosey also has a special and uniquely aggressive relationship with the family dog, Rover (who frequently mutters "ruh-roh" when she comes rolling around to clean up after him), and, in later episodes, Rosey even has a boyfriend and considers marriage.[5] On Mother's Day Rosey wants to participate in the fun, but she gets upset when she realizes she has no memories of her mother. In short, the television series makes it clear that while she may have caster rollers for feet, she is lacking neither a heart nor a mind.

In *The Jetsons* series, Rosey is forty-five years old. If the story is set in the year 2062, that means we should be seeing Roseys in our kitchens by the year 2017. So, are we going to see Rosey mumbling her humble Bronx wisdom as she dusts the living room in the next ten years? Will she push a vacuum? Will she make pancakes? Will she serve tea, held on a platter? Will she have an engine in the rear?

Developments are not far afield from what the writers of *The Jetsons* predicted, with two notable exceptions: First, while all of Rosey's functions exist in robots today, none of them exist in the same robot. Second, all the models today are quite stupid.

Rosey is able to clean, cook, vacuum, and take care of work outside the house as well. She's a general service robot, as the industry term goes.

What's she got? First, like Seiko-chan, Rosey balances on that single wheel, which means she probably has gyroscopic sensors for both axes. Her wheels are attached to her torso, which is connected to her arms, which are connected to her hands—which are actually weird lobster pincers. I don't know what confused designer would give her such brutal claws, but despite this handicap, she can perform all of the household chores and seems pretty strong to boot, since, in the first episode, she's seen shaking a huge dusty carpet from the balcony of the Jetsons' apartment.

I have not been able to discover why Rosey clinks. It might be that she has a loose flywheel or two in her torso to help her pivot on her single wheel. It might be that Elroy's been sticking

5 Season 3, episode 6.

things inside of her. But either way, she is preceded by an interminable clinking and beeping.

With that clinking feature aside, today's robots are able to keep up with the formidable Rosey. In fact, many of them have already far outstripped her in their abilities to perform household chores. But there's something peculiar that Rosey can do that robots today can't touch, even if they do have dexterous hands with an actual sense of touch and a more-perfect BBC-style English accent.

"Would you prefer okonomiyaki sauce or soy sauce?" the robot asks.

In the Kitchen

OKONOMIYAKI IS A THICK JAPANESE GRIDDLE cake that's made out of grated cabbage, egg, flour, meat, and other savories. It's delicious, especially late at night after a drinking binge, around two or three in the morning, when most normal people like to be home and asleep.

Okonomiyaki Robot, or Motoman-SDA10, is there to answer the call. The robot, built by robotics system integrator Yaskawa, used to work in factories, assembling cars. This is the same robot that made the guest appearance in the factory where the Terminator got squished, but these days he's been reassigned to chef duty. With two arms, each with six joints, the multitalented Motoman-SDA10 is able to position its hands to nearly any necessary location in a 3-D space. This flexibility of movement reduces calculation loads and energy costs, and increases speed. So whether it's flipping pancakes or assembling a car, the robot reduces setup time to position the coordinates that an engineer often needs to do prior to the robot doing its job. Wild, tentacle-like arms make service easier, simpler, and less expensive.

When cooking okonomiyaki, Motoman-SDA10 quickly coats the griddle with cooking oil, in strangely precise looping circles, drops in the batter, dead center, and then stops moving and waits. When the bubbles appear in the batter, Okonomiyaki Robot, spatula in each hand, brusquely flips the cake and then mutters, "Would you prefer okonomiyaki sauce or soy sauce?" The robot has some voice-recognition capability, so it will understand when you answer with one of those two words.

It's not as good at conversation as Rosey, and you need to preprogram each of the moves, but once you've got it down, it'll crank out pancake upon pancake for you and your friends, until everyone is sober enough to go home to bed.

Like Okonomiyaki Robot, there's another robot gaining a bit of a reputation for food prep, and that's the "Chef Robot" of Baba

Iron Works (of Avantec).[6] This rig has a long, hinged arm and a device modeled to look roughly like a human hand, with fingers that can grasp pieces of sushi and place them carefully onto customers' plates. Before using the robot, as with Motoman-SDA10, there needs to be some preparatory work, but once the system is up and running, it's able to make enough sushi for a small army in a matter of minutes.

Future Roseys promise much more than stationary cooking of preprepped food in a repetitive process, however.

Future Roseys promise more than stationary cooking of preprepped food in a repetitive process.

Most recently, in March of 2010, a laundrybot was produced. The project was curiously titled "Cloth Grasp Point Detection based on Multiple-View Geometric Cues with Application to Robotic Towel Folding."[7] The project, based on a Willow Garage operating system, showed how a robot would be able to fold laundry and detect how best to do this with the aid of a green screen behind it to help define light and form. Much like Okonomiyaki Robot and Baba's Chef, we are looking at a system that needs some preprogramming before it can get to work. The essential task gets done, but it's a bit like setting an infant down with a box of toys more than having an independent servant who knows exactly what to do.

The University of Tokyo has spent a few years developing what they call the "Home Assistant Robot." This project is an effort to move closer to Rosey, a general household servant. The little bugger, while nowhere near as attractive as Rosey (and commercially doomed because of it), is 155 centimeters (or about 5 feet) tall, moves about on two wheels, and can "learn" what objects are in front of it. The robot can grasp soft objects, like plastic cups, and if it misses or drops the object, can try again. It can operate a washing machine, mop the floor (and move furniture as it does so), pick up dishes, trays, cups, and bowls, and carry them from the dining table to the kitchen sink. But these functions do require substantial preprogramming.

The Dirty Work

IF YOU WERE TO SLICE OUT THE MANUFACTURING precision of Okonomiyaki Robot / Motoman, bolt that onto Home Assistant Robot's perception and navigation, nail in a little of MURATA BOY's popularity, and slather that mess on top of a single pivoting wheel with gyros like Seiko-chan has, you'd end up with

[6] http://www.avantec.jp/.

[7] http://www.eecs.berkeley.edu/~pabbeel/personal_robotics.html.

Mark Stephen Meadows

The older model Roomba doubled as Carlos's skateboard better than we had expected.

a lobotomized version of Rosey. This dream-team mechanism could perform some major household chores, operate complementary nonrobotic machines (like washing machines), serve meals, pick up around the house, all while rolling around on one wheel. It could *mechanically* do this stuff, but cognitively, the robot—if we can call such an oddity that—would be lost. It wouldn't be able to pick up a plate of food, let alone serve one. Preprogramming is what allows these robots to know where the rice is, where the oil is, and where the stove burners are located. It requires planning and pre-scripted input.

There are hundreds of vacuuming robots.

So recognizing a thing that is not already positioned is a cognitive task, and while Home Assistant can handle some of the perceptive and locomotive tasks that would allow this to happen, much work needs to be done in order to shake a carpet or pick up Elroy at school.

And we still don't have vacuuming checked off.

The largest domestic robot market has spawned what is probably the single most influential robot in the world: a vacuum cleaner. There are more robotic vacuum cleaners on the market than you can shake a mop at. Some of them use electrostatic dust pads;[8] some use standard air vent–based suctions; some use rollers or ball-bearing roller devices, while some use treads or mops. None of them do dishes. None of them have one wheel. None of them can make okonomiyaki—but they sure can suck up dust bunnies.

There are hundreds of these robots. The vacuum named eVac (by Sharper Image) is one example; the Koolvac (by Koolatron) another. Then we have the more colorfully named household vacuums that seem more like engine-part numbers from the American motor industry, such as the RC3000 (Kärcher), the VSR8000 (Siemens), the RV-88 (SungTung), the DC06 (Dyson), the VC-RE70V (Samsung), and the V-R5806KL (LG). Each of these products does essentially the same thing, which is keep track of where they are, look for dust bunnies, herd them into a corner, carry them back to a recharge station, digest said dust bunnies and disgorge them into some receptacle or other, and then wait to recharge. Many of them have USB and Wi-Fi connections, which allow for upgrades, location-specific data, or inputting of floor plans and specific task requirements.

The king of this vacuum robot population is the Roomba, the most famous of robotic vacuum cleaners, and perhaps the most famous of all robots. The Roomba is manufactured by

[8] RoboMaid, invented by Torbjørn Aasen, and distributed in the U.S. by Telebrands.

Mark Stephen Meadows

A view of how Roomba bumps his way along a wall. For more Roomba timelapse see http://signaltheorist.com/?p=91.

iRobot, which claims to have sold two-and-a-half million of these units.

If you set the Roomba down in the middle of the room and flip it on, it will spiral outward a few times and then hit a wall. It will more or less ricochet around the room's perimeters and then follow the walls to build an internal map that it will then use while vacuuming. The battery lasts for about two hours. There's a wall sensor, a bumper (so when it runs into something it knows it), cliff sensors (so it won't fall down the stairs), and, according to iRobot, the system updates its coordinate space about sixty-seven times per second.

In terms of something that works, that is no longer relegated to research labs, and that doesn't need to be preprogrammed or remotely operated, Roomba is probably the closest we've gotten so far to Rosey.

A friend loaned me one to spend some quality time with. At first I thought it was lost under the sofa, but when I looked, I discovered that it was all tangled up in the stereo speaker and telephone wire cables, like some cat with a ball of yarn. When I reached down to free it, the thing scuttled off and ran into the far wall. The algorithms built into the system are there for path finding and navigation, learning the position of sofas and walls, and have nothing to do with gesture recognition, symbolic recognition, or anything as complex as language. Despite that, according to iRobot, about half of the people that own a Roomba name it.

Roomba has a frenetic personality. The day I observed my friend's Roomba, it seemed diabolically anal, scooting around on the floor sucking up crumbs and loose hair. Having it in the room with me made me think of a hyperactive beagle sniffing out some tiny criminal. It was neither quiet nor intelligent (things I think robots should be), but I suppose this is part of its family history, since, like some fish that crawled out of a swamp, Roomba comes from the proud line of automatic pool cleaners of the 1980s.

In 2007 and 2008 something happened to dear little Roomba: iRobot began manufacturing a remote-controlled model, something you could log into from afar. According to the iRobot Web site, this system, called ConnectR,

> *. . . enables real-time virtual visits over the Internet. Equipped with high-quality audio and a video camera, the robot is located on-site in the home of the "host" party. Using a computer keyboard, mouse or joystick, the remote ("visiting") party can drive the robot around and interact with those on-site, virtually participating in activities at home or wherever the device is located . . .*

This is great for abusing your pets when you're not home. It's also great for keeping an eye out for burglars, and it's even better for burglars who are keeping an eye out for you. Most domestic robots have a unique wireless message that is used. Such wireless messages can be found and intercepted, and the audio, video, and control data can be accessed by an unwelcome guest. So if you can log into your Roomba from your computer while you're away, then I can hack into your Roomba while you're home. If your Roomba is equipped with a video camera and you happen to be sitting on the sofa watching television at night, then what's to keep me from rolling around the house and doing a full scope-out? You'll just think your Roomba's cleaning house. I'll know when you're there, where you put the keys, which window is open, where the dog sleeps, and where you stash your spare money when you don't think anyone's watching. Then, while you're out walking the dog, I'll stroll over and help myself to your goodies. Or I'll have one of your other robots do it for me.

If you're more interested in having your domestic robot mop rather than vacuum, there are alternatives, but not many. In 2009 Panasonic demonstrated an autonomous floor-cleaner robot at the Tokyo Fiber Senseware exhibition in Italy. Panasonic's Fukitorimushi (which means, loosely, "wipe-up bug") uses polyester fibers less than 1/7500th the width of a human hair, according to a report in *Slash Gear*. The unit is capable of path finding and uses light to differentiate between floors that are dirty or clean. When it finds some new evidence of a crime, it heads over to rub itself on the floor with its microcloth body and then, like Roomba, docks itself for a recharge. The cloth does need cleaning from time to time, however.

We even have robotic bees.

If you have a swimming pool, sidewalk, or lawn that needs cleaning, sweeping, or mowing, there's a robot that will do it for you. Robots today are being developed to not only mow lawns and clean swimming pools, but also to weed, water, clean, and harvest plants. We have the Danish Casmobot[9] and the robotic gardener at MIT[10] (mounted on top of a Roomba), to name only two of over a dozen.

We even have robotic bees. In 2009 researchers at Harvard were provided with a $10 million National Science Foundation grant to create a swarm of flying robotic bees. The RoboBees are intended to function on low power supplies, remain networked, coordinate their flying with the rest of the swarm, and their swarming with the rest of the hive. Additionally, they are intended to have pollination abilities that work with UV and optical sensors. So here we see new forms of robotics emerging from the house, spilling into the yard, and flying across the fields and prairies of a town near you. It's a far cry from Rosey, but the core concept is the same: service.

Tomio Sugiura, president of Sugiura Kikai Sekkei, which manufactures a vegetable-slicing robot, foresees a robot in every home in the near future. "Nowadays, almost every family has a car. In the near future, every family [will have] a humanoid robot that can help out with various things at home."[11]

Robots (humanoid and otherwise) will likely be owned by many, if not most, families in industrialized nations in the next thirty years. And like bees, they will go far, far beyond the home.

In the Street, City, and Shopping Mall

IN THE JULY 1933 EDITION OF *POPULAR SCIENCE Monthly*,[12] a "robot" was reportedly seen in the London metro. About as tall as a person, but

---

[9] http://www.fieldrobot.dk/.

[10] http://www.csail.mit.edu/.

[11] http://www.reuters.com/article/idUSTRE5591JX20090610.

[12] Vol. 123, #1, edited by Raymond J. Brown.

shaped like a hexagonal cylinder, the device was a replacement for a ticket salesperson. Metro riders (and presumably a few curious on-lookers) would twiddle a dial that would then mechanically click to a position according to prefabricated instructions. This mechanical calculator helped people determine the route, rate, and connecting platform. We probably wouldn't call such a thing a robot these days because we see them everywhere in metros and train stations around the world, but a similar customer-service trend is reappearing in Japan.

In January of 2008 the ATR laboratories[13] of Osaka conducted a series of demonstrations of a humanoid robot named Robovie that was a distant and souped-up cousin of the "robot" from the 1933 London metro. At a local shopping center in Osaka the robot was set up to roll around a relatively large space and pick up behavioral data from shoppers. It used sixteen cameras, range sensors, and radio-frequency identification (RFID) technology to watch about two dozen people at any given time. The robot would keep track of people's locations as they were walking, but if a person stopped for a period of time, and especially if they started turning slowly in circles or looking at a nearby map, helpful little Robovie would roll up and ask in a synthesized but childlike voice, "Are you lost?" If the answer was no, Robovie would suggest additional shops in the mall that it thought merited a visit, then, having bestowed its bit of advertorial advice, it would self-contentedly roll off to pester some other shopper. If the answer was yes, Robovie would ask where the person was headed and point the way.

Though I have seen Robovie in action, I can't tell you how precise its answers to the test were. But I can tell you that if a robot were to tell me where to start shopping, I'd be inclined to give it a kick in the CPU. Robotic advertising in shopping malls may be cute at first, but automation in advertising is definitely not for everyone.

From ATR's perspective, they were doing the same thing that the London metro robot was doing: giving people information and helping them do what they had come there to do. The difference is that the London Metro robot was there to offer exactly the information that was sought after. Robovie, on the other hand, volunteered information. It's a pull-versus-push distinction, and a fine one if you stop to consider a shopping mall populated by robots that follow you, all of them about to run over your heels, all of them sponsored by some local shop, all of them chattering in their childlike electronic voices, all of them simultaneously volunteering information on where you should shop as you scamper into the bathroom to hide, and they follow you in because they are genderless (and they also clean the bathrooms when they are not herding people).

[13] Advanced Telecommunications Research Institute International (more on this in following chapters).

Mark Stephen Meadows

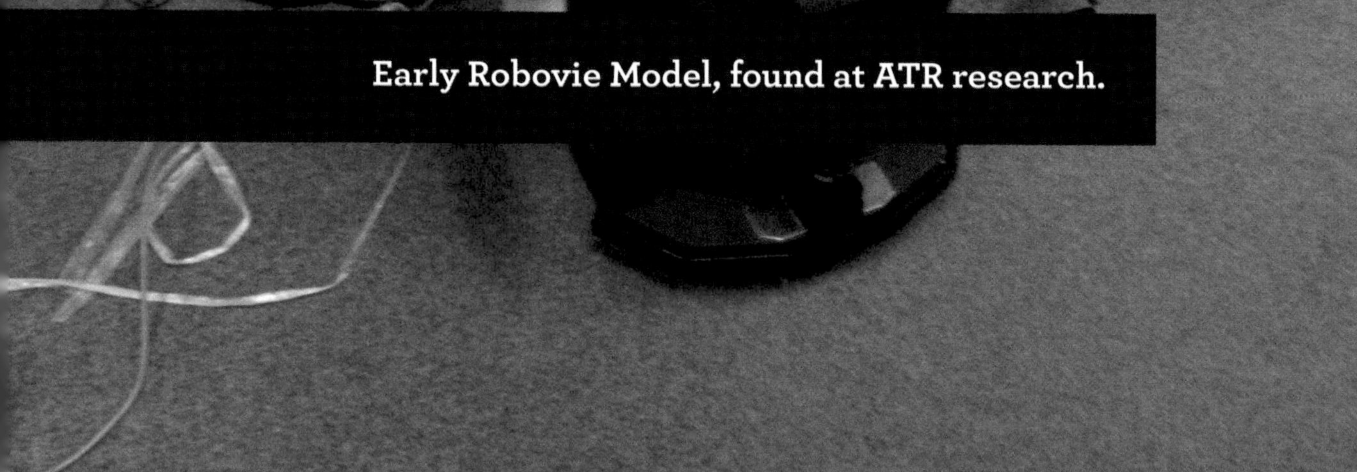

Early Robovie Model, found at ATR research.

ATR, to their credit, considered all this and began to build tools to remedy the problem. First they installed a wireless network which helped the robot communicate with other robots that were also roaming around in the mall. Second, to refine what shoppers wanted and to minimize the need for the robot to offer advice, they installed a system that allowed a shopper to contact the robot via a mobile device before they arrived at the store, entering a list of items they wanted to purchase. Then, when the shopper arrived, the robot would identify the mobile device, roll up to the shopper, grab a little basket, and off they'd go.

Harvesting robots have to see fruit that's hidden behind leaves.

When ATR invited the press to an opening demonstration in the Apita-Seikadai supermarket in Kyoto, the press members all pulled out cameras, and the poor little shopper-robot eventually had some difficulties following the woman it was supposed to be collecting food with because of them.[14] The video of the feeding frenzy can be found on YouTube.[15]

In Fields and Farms

OUT BEHIND THE SHOPPING MALLS OF KYOTO ARE a few remaining fields and farms. Most of the fields are now gone, casualties to progress, but there's a little wind through the branches of an orchard nearby. In a few years robots will be moving up and down between the rows, plucking fruit and pruning branches and trimming back weeds as they go. Vision Robotics Corporation, a San Diego–based firm started in 1999, is developing machines for this work. The job is divided between two robots: one to find the fruit, and the other to grab it. Harvesting robots are a bit tricky because they have to see something (like a cauliflower) that's hidden behind something else (like cauliflower leaves).

The standard recipe for these machines is a manipulator, wheels, and control system. Finding the good stuff without killing the plant is the problem. Some researchers in the UK[16] are working with a combination of radio frequencies, microwaves, and far-infrared spectrography to find things such as cauliflower heads. In the race to replace ourselves, Japan is once again in the lead with robots that collect tomatoes, cucumbers, and strawberries, already functioning in the fields near Kyoto today. They are not collecting fruit from trees, but these ground dwellers are capable of bringing in the goods in a basket at the end of the day, replacing manual laborers each year as they improve in efficiency and increase company profits.

[14] http://robot.watch.impress.co.jp/docs/news/20091214_335825.html.

[15] http://www.youtube.com/watch?v=BckK1EcRA60.

[16] Vegetable Harvesting Systems (http://www.vhsharvesting.co.uk/) in conjunction with the National Physical Laboratories, of Middlesex UK (http://www.npl.co.uk/).

We're far from the land of Rosey the Robot Maid and other simple domestic robots. The romantic notions of a little squeaky household worker with a personality will be replaced by silent-running swarm cyborgs and systems we don't even see that quietly inhabit our houses and yards, noiselessly vacuuming and sweeping, cleaning up the debris we leave behind. The image of a talking, bawdy maid seems less and less likely. After all, talking with servants is not an interaction most people want.

Nonetheless, our list of personal-service robots continues to grow, and outward from the home it goes.

In the waters off the coast of Japan, just down the road from the fields where our cucumber-harvesting robots toil, is a small bay near Osaka. Here, a robot buoy that cleans up offshore oil spills has been known to float from time to time. SOTAB 1 (Spilled Oil Tracking Autonomous Buoy 1) is a 110-kilogram system designed by Naomi Kato, professor of Submersible Robotic Engineering at the University of Osaka. The robot is dropped in the water and then takes on the semiautonomous job of descending about 3 meters underwater, maintaining its buoyancy, tracking the speed of the current via GPS, photographing the severity of the oil spills, and reporting back changes as they happen. When night falls the robot turns off its cameras and continues running footage of what it sees nearby. If it finds some oil overhead, it floats to the surface, navigates with its few small fins, collects water samples, and then sinks again to its previous depth to wait for more until it comes home.

Rosey, clinking and quaint, is nothing compared to the monsters of industry that function in our present-day world. There are six-legged semiautonomous robot vehicles that are taking down trees for lumber companies who want to minimize impact on the forest (stepping over things while cutting other things down being the logic),[17] and we now have robots diving underwater to harvest trees that were flooded under the nearly 50,000 odd dams that were created during the great heyday of dam creation in the 1980s.[18] These robots are able to harvest some of the last old-growth timber in the world, since, being underwater, dams made it a little harder to get to. Not for this saw-toothed submariner lumberjack.

There are also massive mining-industry robots, such as Rio Tinto's Komatsu 930E-4, an autonomous truck, or the largest and most invasive of them all, an autonomous offshore drilling platform.[19] As these things rumble and groan in the mountains and underwater, overhead are self-guided satellites that are designed to attack other satellites.[20] And on Mars, there

[17] Six-Legged Lumberjack developed by Plustech Oy and John Deere.

[18] Triton's Sawfish Underwater Harvester.

[19] SINTEF and Norwegian energy company, StatoilHydro.

[20] India's ballistic missile defense program, primarily designed to knock out Pakistani missiles.

Mark Stephen Meadows

A PePaRo model at TEPIA's Advanced Technology Exhibit, Tokyo.

are a couple of robots crawling around taking soil samples.

We can imagine Rosey spinning on her one wheel, looking for the next carpet to vacuum, as huge robots stride past her into the future of today. Poor Elroy.

But Rosey has the most difficult job of any of these robots.

Child Care, Legislation, and Safety

TAKING CARE OF THE KIDS WAS ALWAYS A BIG PART of Rosey's job, but since today's actual robots lack the artificial intelligence that would allow them to make decisions, surveillance coupled with entertainment is replacing babysitting when it comes to robots' child-care responsibilities. The following systems, all designed for child care around the home, seem to fall in line with much of what is needed for elder care as well.

Roboid, part of an experimental pilot project by the Korean Ministry of Information and Communication (MIC), comes in three different sizes and themes (such as the 16-centimeter Panda, the 22-centimeter Cub, or the 27-centimeter Donkey). These three cuties can read fairy tales to kids or help in foreign-language study. They can tune in radio signals and read e-mail, and they are sensitive to both voice and gesture. A model costs roughly 500,000 Korean won, or about $500.

This type of home robot will be connected to the Internet; some of them, especially in Korea, already are. The Korean MIC has been running feasibility tests with robots designed for home usage in Seoul and its vicinity since 2005, specifically with the Internet in mind. Korea did not want to try to compete with the Japanese advances in bipedal locomotion, but instead, they are sticking with what Korea does best: Internet infrastructure. About three-quarters of all South Korean households already have broadband Internet, so MIC intends to use this infrastructure to handle the processing, sorting, and much of the decision-making that Japan's robots carry around locally. A domestic robot has a lot of brain to handle, and the Koreans are wisely keeping that outside of the physical body of the robot and using their existing network infrastructure to mount the next advance in digital technologies. This is done to push Korean robotics into competition with the Japanese, mostly by cutting unit costs.

"In a nutshell, URC robots just provide hardware with the ability of action while most software comes from the broadband through the wireless Internet. That is the secret to how robot prices can go south," said Oh Sang-rok, the MIC project manager in charge of the URC scheme. The project works with more than 1,000 researchers and over thirty companies.[21]

One of the more interesting projects that the MIC has come up with is a small (52-centimeter) five-wheeled robot called Jupiter. It's able to recognize its owner's voice, and will

[21] For more, see "A Robot in Every Home by 2020, South Korea Says" by Stefan Lovgren, *National Geographic News*, September 2006.

Mark Stephen Meadows

Baby Barnyard Animal toy robots. The chick chirps, flaps its wings, and demands attention. Surveillance not yet included.

go online to search for things like local movie listings or to download music. It has a built-in alarm so that in the morning it can wake the family up. While the family's away or asleep at night, it guards the house. If something suspicious happens, it can call you on your cell phone to let you know, uploading real-time video of what's going on. It costs roughly 1.5 million Korean won, or around $1,500.

Robots that are taking care of our kids, preparing our meals, and driving our cars might need some guidance.

At the Aichi Expo in 2005, two porter-robots that were holding doors misfired and one conference-goer was spanked by the door. The other had his heel grazed (which seems a likely scenario to me, surprised as I was by a mere taxi door). Though these conference-goers weren't hurt, they were vocal, and as a

Mark Stevens Meadows

A FurReal Friends toy robot for kids (the model does not include claws or teeth).

result, industry and government officials sat up and started to point fingers at the possibility of robot-inflicted injuries.

With more robots making appearances in public settings, such as at the shopping malls of Osaka, the risk of something going wrong increases. Eventually a human will be killed by a robot, the media will go completely nuts, and legislation will get bolted into place quite quickly. So several politicians in Japan are arguing that if they legislate guidelines now, it would not only provide a greater public sense of safety, but would also afford them the chance to further expand their lead in the various fields of robotics.

Talks are in the works. Tokyo's Robot Business Promotion Council has taken on this responsibility, setting up working groups to study specific industries, such as medical and elder care, where robots are most likely to slip up. The Robot Business Promotion Council is a subset of the Japan Robot Organization, which has listed several measures they think should be examined. They seem to have started with the obvious. Under the topic of collision with humans, built-in sensors and built-in speed limits have been discussed. These talks branched out into wider topics, and it was soon realized that specific information and knowledge was needed not so much about robots as the conditions they were working in. Currently, different ministries are in charge of setting the regulations that concern the civic areas they cover, and the robots that work for them. Robots working on highways? The Ministry of Land Infrastructure would propose any legislation that would cover them. Legislation would also be determined by the national police force, since they patrol the highways. Robots that are operated remotely, via any kind of radio wave—such as 802.11-whatever or Bluetooth or cell phone—would fall under the scrutiny of the Ministry of Internal Affairs and Communications. Robots that populate senior citizen homes, hospitals, and retirement centers would be examined by the Ministry of Health, Welfare, and Labor.

The International Standards Organization (ISO) has gotten news of these mutterings and has proposed that an international standard for robots be passed in 2011. Japan, Korea, and Britain are all part of the panel involved in creating this set of standards, with Japan proposing to lead the talks. The various Japanese ministries that report back will offer insight, which will then be set up by the ISO and followed (or not) locally.

The Robot Maid, Part 2

WHAT ROSEY BROUGHT WITH HER WHEN SHE ENtered the homes of America's television viewers was not the idea of a machine doing chores. Home-help appliances were already a massive industry. Electric dishwashers, washing machines, blenders, grinders, choppers, meat slicers, irons, and even electric napkin folders had ramped up in popularity in the 1950s, the decade just before *The Jetsons* aired. So the idea of a machine helping around the house wasn't new or surprising.

What Rosey introduced was the idea of a machine that could talk and work at the same time. Rosey introduced us to the possibility of an appliance with personality. This was the most important facet; Rosey had, as she called it, "smahts."

The dream of machine intelligence was what Rosey had, and it's what robots today lack. The American population of 1963 was intrigued by the idea of a robot that greeted you at the door, not one that sniffed for dirt under your sofa.

This intelligence was, and remains today, Rosey's killer app.

Chapter Three: 2001: A Space Odyssey

On Robots That Play Chess, Tell Jokes, and Invest Your Money—Intelligence, Part 1

> We have probably developed our aggressive instincts through evolution and being in a dangerous environment, so machines wouldn't have that background, so perhaps they wouldn't be aggressive and couldn't be malevolent, unless they're deliberately programmed by us to be malevolent or aggressive, which incidentally is what many of our machines are programmed to do right now.
>
> —Arthur C. Clarke, foreword to *HAL's Legacy: 2001's Computer as Dream and Reality*, edited by David G. Stork

SOME ROBOTS SCUTTLE AROUND ON THE FLOOR and get tangled in the wires under the sofa, while other robots actually live in those wires and circulate in the cables that connect our houses. These robots are like electricity, and move between networks like ghosts, composed only of bits of electricity or software. These are virtual robots.

A robot doesn't need a body to get a job done.

Imagine a robot operating system that is not attached to Rosey or the Terminator but is still capable of controlling the physical assembly. The robot's operating system would contain all of the personality, goals, and interfaces for controlling the information that triggers the servos and drives the pumps. Or imagine a software system that is simpler, even, and guides missiles, or predicts the stock exchange.

These systems are also called *robots* because a worker (or robot) can perform physical work (with their hands, for example) or knowledge work (with their mind).

The term *knowledge worker* was coined in the late 1950s by Peter Drucker, an Austrian author and consultant. He used the term to describe someone whose work is based on interpreting and redirecting information.[1] A knowledge worker's job is to route information in a manner that helps people, improves efficiency, or creates new ideas (among other things). Knowledge workers do not need to be physically located in any specific place to get their job done because they are working with knowledge. Examples of knowledge workers include lawyers, teachers, journalists, doctors, designers, and engineers.

---

[1] Linda Stone, a writer who has worked at both Apple and Microsoft, has neatly taken us into the twenty-first century with the term *understanding worker*. I find it fits quite nicely here. She says, "Knowledge becomes understanding when related to other knowledge in a manner useful in anticipating, judging, and acting." This is to say that robots will soon be able to anticipate, judge, and act on those predictions (which we will explore further).

Knowledge Workers

ONCE UPON A TIME, IN THE YEAR 2000, I WORKED as a consultant for a division of Oracle, a software company based in Redwood City, California. My job was to design a conversational system named Alan. It was intended to provide that all-important customer service—to answer questions, give a little advice, and generally replace the people that had been doing this in Bangalore, the unfortunate folks whose job it was to traverse a tree of probable responses and give polite replies. At the time I was more motivated to help those poor customer-support folks than anyone else. At least, that's what I told myself. I often took the absurd yet motivational route of imagining myself as the savior of some poor woman, sitting alone at her desk with a phone clamped to her head, listening to some angry customer abuse her from thousands of miles away while she is forced, at risk of losing her job, to give prescripted answers that she reads from a piece of paper that is glued to the wall of her cubicle. This poor woman, I told myself, was being roboticized. I would save her soul, even if it cost her the job she perhaps desperately needed. The dream, at least, kept me well-mannered and in my seat.

A knowledge worker is someone whose work is based on interpreting and redirecting information.

In proposing the project it seemed that most of the chatbot's system could be assembled pretty easily. First, we'd invent some questions a customer might ask and slot those into one of a number of categories. Then we'd invent some answers that a service representative might answer and slot those into their categories; finally, when a customer asked one of the questions, we'd just shoot back the answer. At least, that was the theory. It did the job for a year or so. Though Oracle saved some money on their customer relationship management (CRM) systems, and the engineering team didn't have to work weekends, I never met my imagined Indian princess. What did happen, however, was that Alan needed updating, and rather than change what the robot said, they pulled the plug and killed my first chatbot.

There have been thousands of these kinds of robots, perhaps tens of thousands, produced in the last decade. The primitive, ham-fisted approach above has been cultivated into a subindustry of thousands of people creating complicated forms of interactive narratives that are able to maintain awkward, if not horrible, dialogue. The list is long, and players flicker in and out every year. They include Artificial Technology GmbH, Media Semantics, Verbots, SitePal, CyberTwin, Oddcast, Artificial

Solutions (formerly KiwiLogic), Daden Limited, PikkuBot, Sine Wave Actorbots, MindMentor, and many, many others.[2]

The ALICE[3] project is a system that many people have groomed and enhanced. The system requires massive amounts of text and typing by many people so that it has a level of believability and humanity.[4] The project's inventor, Richard Wallace, believes that when there is enough volume, the value will be clear. In other words, he believes that the larger the quantity of possible responses, the higher the quality of the interaction. Though I agree with this to a certain extent, he goes too far when he says that his system is, essentially, how a human works.[5]

Each of these projects tries to create the most humanlike chatbot possible. Unfortunately, it seems to me that the Microsoft paper clip has been reincarnated with the face of a corporate and very arguably sexy he-woman. There are chatbots from offices that are conducting a campaign of disturbing and horrifying us in progressively snottier ways.

Dialogues with a chatbot, when found in the wild, are horrific because there's no context for the conversation. Context (that is, the reason for the conversation, the subject that's agreed upon, and the goals of the conversation) is the important part. If I'm walking down the street, for example, and someone comes up to me and says, "Hi!" and stands there and stares at me, I'll most likely just keep walking. It's the same if I see one on a Web page.

Despite the rough design beginnings (all interfaces improve over the years, after all), there are some projects that seem to be doing well, provided they are implemented in context. Elzware,[6] based in the UK, is a company that has been building corporate chatbots since 2002. Here is a typical transcript of a conversation between Yhaken and a customer ("User" is the human):

Yhaken: Hi, there. What can I help you with?

User: Can you talk to me?

Yhaken: I do not have an audio output at this time, but I am more than capable of talking to you. What do you want to know?

User: Okay. What can you do?

Yhaken: I am capable of cross- and up-selling a client in line with an understanding of their behavior patterns. Like a trained human being.

[2] Their Web sites are Artificial Technology GmbH (www.artificial-technology.com), Media Semantics (www.mediasemantics.com), Verbots (www.verbots.com), SitePal (www.sitepal.com), CyberTwin (http://site.mycybertwin.com), Oddcast (www.oddcast.com), Artificial Solutions (formerly KiwiLogic) (www.artificial-solutions.com), Daden Limited (www.daden.co.uk), PikkuBot (www.pikkubot.de), Sine Wave Actorbots (www.sinewaverobots.com), and MindMentor (www.mindmentor.com).

[3] ALICE is an acronym for Artificial Linguistic Internet Computer Entity.

[4] There is a tremendous amount of material on the Internet about ALICE. A good starting point is http://alicebot.blogspot.com/.

[5] If that were the case, we'd have it working like a human by now.

[6] Elzware does a fine job of creating automated CRM services today far beyond what we were imagining at Oracle (http://elzware.com).

Illustration by Elzware

Elzware's idea of what Yhaken might look like.

Once upon a time, a real human used to spend time on a real phone and talk with real customers. Unfortunately, CRM managers gradually automated the work their employees did to such rote and automatic tasks that it only made sense to replace those same employees with equally rote and automatic machines. These poor people were sent off to the scrap heaps in favor of an assembly-line approach for paying customers. Just as workers in Detroit's automotive factories were replaced with robots, CRM staff were replaced with chatbots.

Customer relationship management is becoming a robotics subindustry. In fact, today about three-quarters of the 800 numbers in the United States either currently use or are in the process of adopting speech recognition systems.[7] The American 800 number system is becoming a virtual world, an auditory landscape populated with robots. According to Forrester Research, by 2010 spending on CRM systems surpassed $11 billion annually.[8] There are phone robots that will help you get a new American Express card, file your AT&T payment, order a shirt via Banana Republic, buy a book at Barnes & Noble, schedule a rental time with Dollar Rent-a-Car, find your FedEx package, tell you where to grab a Greyhound Bus, price a hammer (and sell it to you, physically) at Home Depot, and take your article submissions at the *New York Times*. Walgreens, Wal-Mart, Walt Disney, and U.S. Homeland Security all now funnel more than 90 percent of their public relations through robots.

You can use these instead: http://www.dialahuman.com/ and http://gethuman.com/.

We're Past 2001 . . . Where's HAL 9000?

IN 1968 ARTHUR C. CLARKE PREDICTED[9] THAT RObots would be indistinguishable from humans in just a few decades. He predicted we would have a computer that could talk, feel emotions, and give advice, and that it would appear before 2001. Now, in 2010, it isn't yet visible.

The film *2001: A Space Odyssey* was written by Stanley Kubrick and Arthur C. Clarke, two great minds of fiction and storytelling. The

[7] Database Systems Corporation also includes this interesting statistic: "Companies that 'selectively force callers through the IVR [interactive voice response]' achieve the highest IVR success rates. Forty percent of participants selectively force callers through the IVR for some services. These same companies averaged the highest IVR Success Rates—more than 40 percent (percent of IVR completed calls versus total handled calls). Conversely, companies that present IVR services as 'optional' average an IVR Success Rate of only 13 percent."

[8] "CRM Best Practices Adoption" by William Band, Sharyn Leaver, and Mary Ann Rogan (January 10, 2008).

[9] Crevier, Daniel, *AI: The Tumultuous Search for Artificial Intelligence*, New York: BasicBooks, 1993.

MGM / The Kobal Collection

Inside HAL's mind.

United States Library of Congress gave the film big prestige, calling it "culturally, historically, or aesthetically significant." It received four Academy Awards, four BAFTA awards, and a Hugo. It was selected by the American Film Institute as one of the top 100 greatest American films of all time.[10] The National Film Registry chose it for preservation, in 1991.

The film features a peculiar robot named HAL 9000, a classic example of Strong Artificial Intelligence (Strong AI), and a good example of humanlike AI.[11]

In the film's year of 2001, HAL is the latest product in machine intelligence which can reproduce (or at least mimic) functions of the human brain. HAL is considered the brain and central nervous system of a spaceship headed to Jupiter, and the 9000 series is, as we hear in a newscast, "the most reliable computer ever made." As HAL himself puts it, "No 9000 computer has ever made a mistake or distorted information. We are all, by any practical definition of the words, fool-proof and incapable of error." HAL is never frustrated, despite working with humans that are so much less intelligent than he, and HAL says he enjoys his working relationship with the astronauts. Whether he has real feelings isn't something anyone can truthfully answer, partly because his face is a single, red, cyclopean eye.

Everyone aboard the good ship *Discovery One*, headed to Jupiter, seems to be a happy little family, based on a strong foundation of trust. HAL is capable not only of language and image

10. It was listed as number twenty-two, after 1940's *The Grapes of Wrath*, and before 1941's *The Maltese Falcon*.

11. "Strong AI" is different from things like bots or "applied AI," in which machines perform a specific task in a specific context. "Strong AI" is about a system that more or less (well, more) behaves like a human.

processing, art appreciation and speech recognition, but he can also play chess, tell lies, keep secrets, read lips, and, best of all, kill people.

How much of this do we have today?

A Robot Walks into a Bar and Asks for a Screwdriver

A SYSTEM LIKE HAL IS ABOUT AS FAR AWAY AS JUPITER. In order for robots to reach the dream that science fiction has presented of a humanlike AI, we're going to have to jump what I think is one helluva technological barrier. The goal is to create a system that's able to contextualize random information, recontextualize it, predict, adapt, lump things together into general categories, maintain contradictions, and understand why. Common sense, cause and effect, logic, belief, and the ability to reason (whatever these things are, and however they relate to one another) have to then be wired into other abstract notions such as understandings of other, self, subject, object, and a sense of those ever-elusive concepts known as Time and Space.[12]

"When will robots be as smart as humans?"

Then, if we can take three of these AI systems, and they can sit around in a bar and make jokes, we're in business. That's my on-the-record goal for humanlike AI. That's the Meadovian Metric: three AI systems that are able to joke around.[13]

This won't happen anytime soon. And in fact, it doesn't need to happen at all.

It's on everyone's mind. "When will robots be as smart as humans?" "When will artificial intelligence arrive?" "What is conscious?" and "When will we be able to know that humanlike AI has truly arrived?"

These are vague questions, and the answers are often even vaguer.

In 1965 Gordon E. Moore, Intel's cofounder, developed what was either a great stroke of engineering or of marketing.[14] He wrote Moore's Law. This theorem points out that processor speeds double approximately every two years. Around the year 2001, $1,000 purchased 10^7 calculations per second, which is about what an Apple IIE is capable of performing. Because of improvements in processor speed, $1,000 processors will begin to produce calculations of 10^{15} sometime around 2030. This means

12 It is difficult to find a major philosophical work that doesn't, at some point, get tangled up in defining these issues. Regardless, if ever philosophy needed to be pragmatic, it is now.

13 I'm being serious, because this is for a test of *human*, or at least *humanlike*, intelligence. At the risk of getting stuck in the weeds of technical details, because I'm soon going to argue that this is a waste of time, the point here is that three systems interacting passes "the complexity barrier" both socially and contextually. Telling jokes means that the system can shift between various contexts, semantics, and ontologies in inventive ways. And they can't be stuffy, please, which HAL obviously is.

14 If it was engineering, it was prescient. If it was marketing, it was even more prescient, because it encouraged computer users to never be satisfied with what they had, to know they could always get more, and to throw out their old purchase after only a couple of years. If Ford had thought the same way, we'd all have faster cars.

that we can project how many processors and how many calculations we have running in the world during any given year, and how many we are expected to have.

Moore's Law says that this doubling of processor speeds will continue, and new technologies that are appearing support that. One new technology is called "memristors." Though they have been in development for nearly three decades, in 2008, HP Labs announced the development of "switching memristors," an electronic component that records the value of the current flowing through it after the current has stopped. In essence, it is a system for calculating and storing data at the same time. Nowadays researchers at the University of Michigan are using memristors to build a system for learning and memory that's modeled after a cat's brain.

"We are building a computer in the same way that nature builds a brain," said Wei Lu, an assistant professor in the University of Michigan's Department of Electrical Engineering and Computer Science.[15] "The idea is to use a completely different paradigm compared to conventional computers. The cat brain sets a realistic goal because it is much simpler than a human brain, but still extremely difficult to replicate in complexity and efficiency . . ."

Lu has connected two electronic circuits with a single memristor and has demonstrated that this system is capable of what's called "spike timing dependent plasticity." This means that the connections between neurons are able to become stronger when they're both stimulated. "Spike timing dependent plasticity" is considered to be the basis for memory and learning in mammalian brains, as this is the kind of activity that establishes neural pathways and reinforces lessons as memory. This offers big promises for developing forms of artificial intelligence because it solves many problems that software has had to address. So it seems that Moore's Law is in no danger of being broken by laggy hardware chips.

Moore's model lends itself well to extrapolation.

Moore's model lends itself well to extrapolation.

Kevin Kelly, in his interesting essay, "Dimensions of the One Machine," equates neurons with transistors, and points out that we have "approximately 1.2 billion personal computers, 2.7 billion cell phones, 1.3 billion land phones, 27 million data servers, and 80 million wireless PDAs," which is roughly "five orders more transistors than you have neurons in your head." Kelly points out that by sometime between 2020 and 2040, the "One Machine" will surpass all humanity's processing power.[16] As far as I can tell, Kelly

[15] *TechNewsDaily*, "Cat Brain Inspires Computers of the Future," by Charles Q. Choi, April 16, 2010.
[16] Kevin Kelly's article is at http://www.kk.org/thetechnium/archives/2007/11/dimensions_of_t.php.

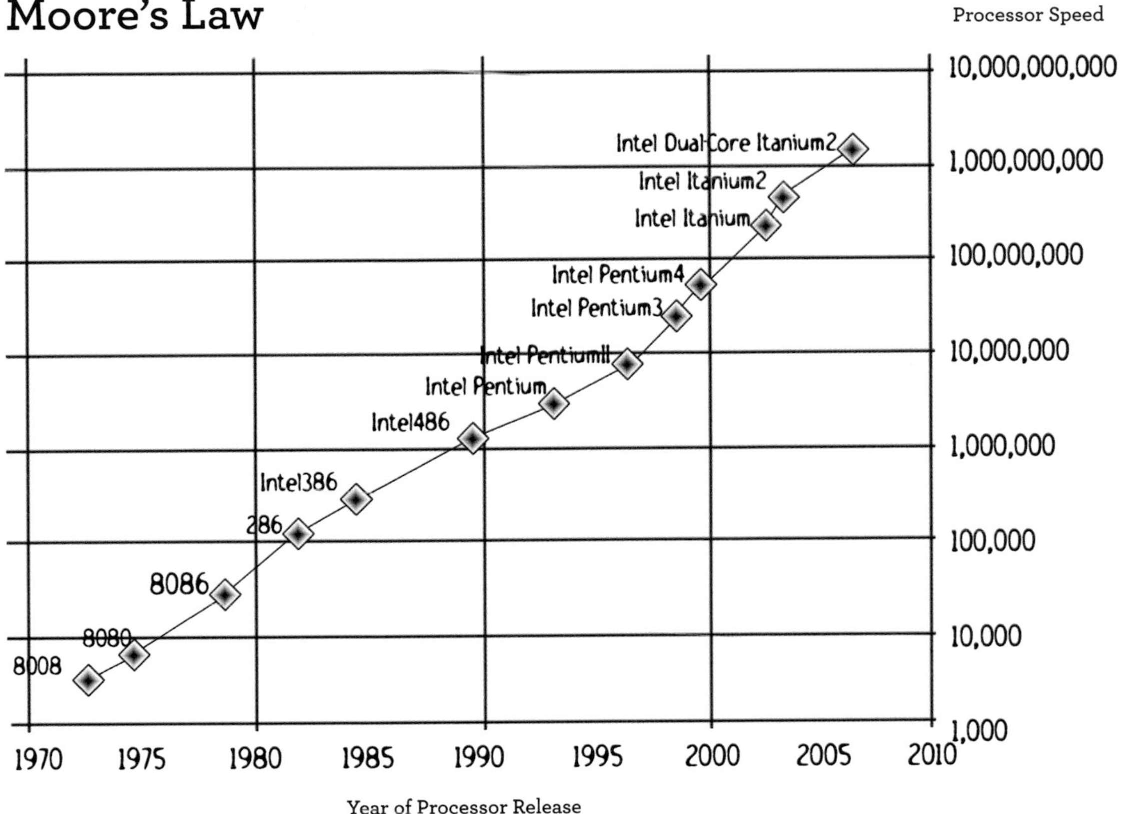

stops short of making claims that it will become self-aware, or transform into a humanlike AI.

Ray Kurzweil, on the other hand, exercises no such restraint.

Author, inventor, and futurist Kurzweil claims, via an argument called "Accelerating Returns," that we will arrive, for the same reasons and at about the same time as Kelly's giant processor collection, at a machine that is intellectually indistinguishable from a human. Kurzweil's argument differs from Kelly's because Kurzweil is looking now at the application (intelligence), not just the machine (processors).

Kurzweil starts along the same lines as Moore and Kelly: In 2001, $1,000 purchased 10^7 calculations per second, which is about what an ant is capable of, and processors will begin to produce calculations of 10^{15} sometime around 2030. Kurzweil also claims that we can measure intelligence by the number and speed of processors. Not just computing power or neuron count, but intelligence. So by 2045, $1,000 will buy you intelligence more powerful than all of the human race.

What's more, as Kurzweil and many others argue, this will create something called a

Singularity. A term popularized by science fiction author Vernor Vinge,[17] a Singularity is a moment in human history when machines will become so smart, so inventive, that they will be able to build themselves, and the human era will end. Like the dinosaurs, our days are numbered.

These are very exciting arguments. More transistors equals more intelligence. With more intelligence comes yet more intelligence. That intelligence will be so smart that we'll be fossilized and a new era will dawn, flying from our hands, and populating the entire universe with its scintillating nanotech intelligence. Some people say this revolution will "wake up the universe," as all matter is gradually transformed into thinking mechanical computers composed of hustling nanobots, all of them crawling to re-form the universe in our image of the Noosphere. The theory says that we live in a time when all is in pivot, when all is potential. And according to this argument, all of us living today will see this new era dawn.

The "Accelerating Returns" argument is peculiar. Aside from being subtly misanthropic, there are two problems with it. The first is that it assumes intelligence and human evolution to be quantitative and mechanical. This doesn't seem right, because human evolution and intelligence may rely on quantities and mechanics, but intelligence and evolution are not defined by them. More important, however, is the fact that I don't think we even know what intelligence is, and so measuring it seems misguided.

Let's go back to ants as an example. Ants compose the largest biomass on Earth. They take up about one-fifth of the total biomass on the planet. So if we line up all the ants in the world, and cut off all their little heads, and remove all their little brains, and put them all in a row and measure them, we would have more neural connections than all the humans in the world.[18] So if intelligence is simply stocking and stacking, then ants should be the ones bulldozing us, not vice versa (if, indeed, one's ability to bulldoze another indicates intelligence).

Bigger isn't better and faster isn't smarter. Ultimately, the "Accelerating Returns" argument is a bit like nineteenth-century phrenology, or craniometry, in which someone's personality was determined by the dimensions of their skull. Just because I have a big head does not mean that I am smart (it might indicate I have encephalitis).[19] Phrenology or craniometry (or any gross quantitative assumption) can give us clues to developmental physiognomy, or the

[17] Vinge did this in *True Names*, and *The Peace Wars*. It must be stated, however, that mathematician I. J. Good first brought this up in 1965; Vinge mainly popularized it.

[18] Encephalization quotients aside (because the individual isn't what we're looking at), ants account for roughly 15 percent of the Earth's animal biomass, and humans for about 14 percent. Proportionally, an ant has about ten times more brain mass than a human does. This means that there is more ant-brain mass than human-brain mass. Also, note that vertebrate brains scale allometrically, rather than isometrically (for invertebrates), with body size. Anyway, if that argument with ants isn't right, take insects in general. And if it's still off, just keep adding other species' brains onto the scale, and we'll get the same result.

[19] Or, more likely, pigheadedness.

psychology of phrenologists, or infections, but quantities like these can't help predict when artificial intelligence will appear. I, for one, prefer to take contemporary methods of approaching old problems with a little salt. Not all truth lives in numbers.

My problem isn't with the argument, but the conclusion.

Defining intelligence, let alone quantifying it, is a slippery job. Few humans, even very intelligent ones, will be able to say what it is. Intelligence quotas have never been thoroughly welcomed as a means of understanding human ability, and the basic effort of measuring intelligence seems to be revised every twenty years. Geniuses have been overlooked by their fellows. We have a hard-enough time recognizing intelligence in other humans. There are many kinds of intelligence, including social, emotional, and artistic (people are often referred to as a *mathematical* genius, or a *musical* genius, which indicates these different kinds of intelligence). There are also many forms of intelligence we still don't understand. Many people have proposed that intelligence isn't entirely contained in the brain in the first place.

Not all truth lives in numbers.

Imagine that we humans make smarter-than-human machines. How can we make something smarter than us if we're all in such disagreement about what *smart* means? Bigger than us, sure. Faster, yes. Quantitative things we can see. However, intelligence seems to be an application of the hardware more than the hardware itself, and so it's hard for me to get past the first sentence of this paragraph if I really consider what it says.

Setting aside the above gripes about quantitative methods, there are other problems with these ideas, such as various "laws" that come into conflict. These laws are about things like "code bloat" and the delicate balancing act between technology, economics, and sociology. Network speeds increase over the years, new technologies arrive, religious fears influence the culture, culture influences technology, and all leave little space for a Strong AI to emerge, even if the "accelerating returns" argument were valid. This is not technology on the order of making a robotic gun; it is technology on the order of making stem cells.

So using quantitative arguments to predict when something that is nonquantitative will arrive, seems quite queer to me. Memristors and nanotechnology hold massive promise (and equally massive threats), but intelligence seems to me to be more application than hardware platform.

I doubt that the Singularity (the theory that smarter-than-human machines will start to make progressively smarter machines) will occur in the way that is commonly imagined today. Machines do not make machines without human guidance. Some instances, such as genetic algorithms,[20] can be found, but it makes sense to look at how history is likely to repeat itself. Consider an industry in which robots are the primary creators of a product better than us—the automotive industry. Cars are made by robots, but these cars and these robots haven't created cars that are creating faster cars. The main thing they've created is pollution. In fact, these robots making cars have created more pollution than they have cars.[21] From where I sit, what we have to look forward to is not superintelligence, but targeted advertising, shopping bots that hassle us, watches that insist we buy video transceivers, iPads that track our reading habits, and shoes that tell us where to go. Things that we won't even identify as intelligent. I don't see demigods striding over the sunny horizon anytime soon, but I do see lots of little devils climbing out of the gutters of valueless commerce.

How could we even recognize humanlike intelligence?

The Singularity (and the industries, such as schools and workshops, that the Singularity is creating) is a kind of religion based in science fiction. It is as if we've made a big tangle of wires called the Internet, and we've had our ear next to it, listening, and we're getting increasingly afraid that we'll hear a voice whispering something malevolent. It would be funny if so many people didn't take it so seriously.

But let's set all this aside and consider that it could happen. How could we recognize it?

Alan Turing, one of the greatest minds to work in computer science, and most probably the inventor of artificial intelligence, has a couple of answers for this question. In 1950 he proposed a test that has come to bear his name: the "Turing Test."

This test proposes that a judge engages in conversation, via text chat, with a human and a machine. If the judge can't tell them apart, the machine is intelligent. Some have argued that this is a test for consciousness. Turing's claim was, "If it appears to be [conscious], it is."[22]

I'm not sure about this, either, and the reason is because, once again, we're encountering a mechanistic point of view. An egg is not an egg just because it appears to be an egg. It's an egg either because a chicken eventually comes out of it or because I eat it, not because it looks like an egg. After all, intelligence is only good if it does something. It can't just look intelligent.

If we set these questions of defining intelligence aside, we have problems with the judging process, too. As an example, let's say that Albert

[20] For example, see the work of Karl Sims and others that looks at locomotion and ways of testing possible variables.

[21] If the average land-based vehicle creates 50 kilograms of pollution each year, and most of these vehicles last for fifteen years, then they make about 2,250 kilograms / 5,000 pounds of pollution while in use—more than the car weighs. Then, the car gets thrown out.

[22] Turing later disputed his own original version of the test, and seemed, finally, to rest as a skeptic.

Einstein—who most would agree was a pretty smart guy—writes a program that's as smart as him. In room #1 he flips his AI program on. Machine humming, he then leaves, enters room #2, and sits down at his own computer. I, meanwhile, have the honor to serve as the judge of all this, so I sit down in room #3 and crack my knuckles as I get ready to engage in what will surely be an amazing test. Which will be Einstein? Which will be his supersmart kinder-bot? My job is to tell them apart, and when I do, to try to identify the robot. But if I can't tell them apart, then the robot passes the test, and we have something that may be conscious.

Staring at the blinking cursor, there is a pause, and the following text appears on the screen:

```
Über einen die Erzeugung und Ver-
wandlung des Lichtes betreffenden
heuristischen Gesichtspunkt!!
```

I don't even know what this means. How the hell am I supposed to tell if this is an intelligent entity that's typing to me? I can't even tell if this is just a computer that's acting like a human being, since my assumptions about culture, psychology, and language skew my judgment. Hell, even my assumptions about assumptions jettison me out of this Schrödingerian box. If Einstein were on that computer and he was typing to me in German, and I'd never even seen German because I grew up in Botswana, am I really the best person to say that this entity is intelligent or self-aware?

The weakest part of defining intelligence is our ability to recognize it when we see it. Most of what I have written I have some doubt in, but I am sure that I, at least, should not be a judge of it.

Should anyone?

01110100011010000110100101110011001
00000011010010111001100100000001101
001011011100111010001100101011011000
1101100011010010110011101100101011011
10011000110110010100101110[23]

And, anyway, Einstein had difficulties with language when he was a kid.

So, in order to identify intelligence in a machine, we'll need to dump our anthropocentric, or pre-Copernican, definitions of intelligence. This pre-Copernican view of intelligence assigns humanity's intelligence as the arbitrary center of a galaxy of possibilities. Human intelligence is the measuring stick, the unit of measurement, and the eye that measures. We consider ourselves the pinnacle of evolution when it comes to all things intelligent. We paradoxically see ourselves as the pinnacle of intelligence on the face of a planet where we are only just beginning to discover how species as simple as bacteria[24] are able to

[23] "This is intelligence."
[24] See Bonnie Bassler's work on bacteria-to-bacteria communications (Princeton).

communicate, let alone things like whales or chimpanzees.

A few years ago I met a programmer who lived in the Black Forest in Germany. One day he found a baby crow and raised the thing as a pet. It loved dog food, and the programmer's dog hated that it loved dog food, but neither the crow nor my friend cared what the dog thought. The dog was left to defend his own food. The crow, my friend noticed one day, would sit on the windowsill and make meowing noises, like a cat. At first my friend, who I still to this day think of as a pretty intelligent person, didn't understand why, until after about a week of this behavior he noticed what was up.

Intelligence seems to be related to the ability to adapt in unpredictable circumstances.

When the crow would make the meowing noise on the windowsill, the dog would run outside to look for a cat, at which point the crow would hop down off the window and eat the dog's food. The crow was just a little smarter than the dog.

My friend told me that after the bird had collected the dog food, it would hide it. But if the crow noticed him watching while it hid the food, the crow would then later rehide it. This crow was thinking, "I know you know how this works, and because you're watching me bury this, you obviously think it's valuable, so I'm going to predict your behavior and bury my treasure somewhere else."

Crows (more specifically, corvids, including crows, ravens, magpies, and nutcrackers, to name a few) have surprised more than a few observational scientists in recent years. They've shown broad abilities of memory, employing complex social skills, reasoning powers, and perhaps most impressively, a striking ability to not only use, but also create, tools. They recognize individuals (both humans and other corvids), and are able to mimic sounds other animals make to fool animals (and humans) into doing things that gain them rewards, similar to my friend's experience.[25] Similar examples of "intelligence" can be found throughout the animal kingdom.

To say that a monkey isn't as smart as us but smarter than a crow, just because a monkey thinks more like us than a crow, makes about as much sense as saying that a monkey is more beautiful than a hummingbird for the same reasons. It's patronizing.

Intelligence seems to me to be related to the ability to adapt in unpredictable circumstances. It involves prediction, memory, analysis, and—the hardest of all—the ability to self-regulate and self-change. Being able to predict means that a system has a memory, and a sense of cause and effect, but simply predicting things isn't enough; the ability to change to conform to another system—and to do it in a circumstance

[25] See the work of Christopher Bird, Cambridge University.

that one has little control over, or knowledge of—seems more like intelligent behavior to me. And bacteria can do just that.

So the question seems open, the problems of Strong AI not yet understood, and a solution for creating one as far away as Jupiter. In sum, my hunch is that our emotions probably have more to do with our intelligence than our logic, but I have a hard time rationalizing this passion and putting it into words.

A Cyclops Sees Deeply

AI IS NOT DANGEROUS BECAUSE IT MIGHT GO BERserk; it's dangerous because we blindly trust it.

In *2001: A Space Odyssey,* like every good robot movie (from the West, at least), something goes horribly wrong. HAL 9000 turns from a mild-mannered smooth talker into a cyclopean Beholder.[26] The initial system tremble, the early-warning signs of instability in their interaction, happens during a game of chess between HAL and the ship's captain, Dave. A careful observer will see that there is an ill-declared move. HAL deceives Dave, but Dave doesn't notice. He doesn't notice because he trusts HAL implicitly.[27] This is HAL's first test. Then, following a curiously intimate man-machine moment of art appreciation, HAL asks Dave a "personal question." HAL asks if Dave is having some "second thoughts" about the mission. This is HAL's second test, to see if Dave is capable of performing the exceptional duties he has in front of him.[28] Dave fails again (at least in HAL's eye), and the interaction between them starts to really wobble. The trust that the robot has for Dave, and the trust that Dave had for HAL, is seriously shaken.

The system ruptures when HAL makes some strange judgment calls, and appears to be malfunctioning. When Dave asks HAL what's up, he says it's human error. But this just doesn't seem right, and so Dave and Frank consider disconnecting HAL. Of course, HAL overhears all of this and the situation throws HAL into loops of logic. At this point HAL is clear that Dave and Frank will fail at their mission, but he

---

[26] I can't help but see HAL (when he attacks Frank) as a two-armed, one-eyed Beholder from the *Dungeons & Dragons Monster Manual, First Version.*

[27] It is a replaying of a very famous game—Roesch vs. Schlage, Hamburg, 1910.

[28] HAL: Well, forgive me for being so inquisitive, but during the past few weeks, I've wondered whether you might be having some second thoughts about the mission.
Dave: How do you mean?
HAL: Well, it's rather difficult to define. Perhaps I'm just projecting my own concern about it. I know I've never completely freed myself of the suspicion that there are some extremely odd things about this mission. I'm sure you'll agree there's some truth in what I say.
Dave: Well, I don't know. That's rather a difficult question to answer.
HAL: You don't mind talking about it, do you, Dave?
Dave: No, not at all.
HAL: Well, certainly no one could have been unaware of the very strange stories floating around before we left. Rumors about something being dug up on the moon. I never gave these stories much credence. But particularly in view of some of the other things that have happened, I find them difficult to put out of my mind. For instance, the way all our preparations were kept under such tight security and the melodramatic touch of putting Dr.'s Hunter, Kimball, and Kaminsky aboard, already in hibernation after four months of separate training on their own.
Dave: You're working up your crew psychology report.
HAL: Of course I am. Sorry about this. I know it's a bit silly. Just a moment . . . Just a moment . . .

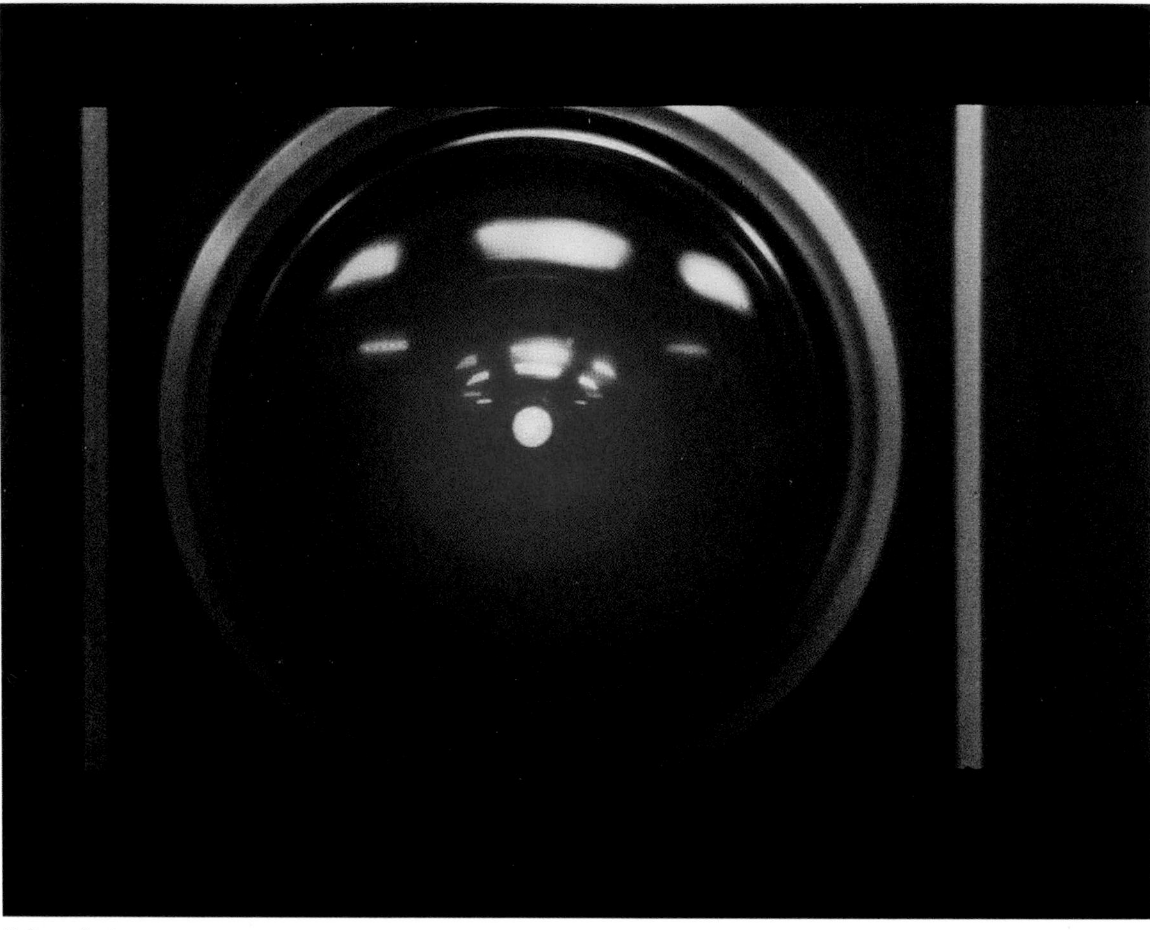
MGM / The Kobal Collection

Hal's cyclopian eye.

also feels he has to protect them. But if he protects them, the mission will fail. Second, HAL was instructed to keep their mission secret, and to do it without distorting information. This was also a bit of a contradiction that he was unable to resolve. These pairings caused a rupture in the algorithm named HAL 9000, and ultimately led to disaster for the company and crew.

The failure isn't a result of the humanlike robot, but the robotlike men. The astronauts, chosen for their machinelike dependability, are more robot than HAL. HAL, unique in the film for his emotional and graceful demeanor, is more human than Dave and Frank. HAL isn't evil, nor is he malfunctioning. In fact, HAL is forced, by his human designers back on Earth (because of decisions they made a long time ago) to behave as he must. But HAL's human designers never anticipated such circumstances, and so the system wobble is a result of poor foresight by mechanically-minded people, not a result of evil by a human-minded machine. HAL becomes a diabolical killer because of unforeseen design decisions. It is, as he said, "human error," on both the part of Dave and on the part of the humans that created HAL. This is what causes him to kill the astronauts.

HAL kills the crew because he has been indirectly instructed to do so.

HAL is a remote (and future) version of human designers' decisions.

HAL kills the crew because he has been indirectly instructed to do so by his designers and programmers, who didn't consider the unintended consequences of their instructions. This is what makes HAL the most human of all the members of the crew. If we think of HAL as a system like those in Asimov's *I, Robot,* we can understand HAL to be an utterly, unwaveringly, 100 percent faithful and inflexible missionary. His mission is to make sure that the ship arrives at Jupiter's moon with the secret of their mission intact. He is forced by conflicting logic, not foreseen by his designers, to kill off people who might discover this secret.

The film *2001: A Space Odyssey,* as well as being about weapons, intelligence, and human evolution, is ultimately about logical decisions. It is very similar to Asimov's story, "Runaround." These stories are about how logic is not always reasonable, and this presents a very poignant angle on humanity and robots. After all, it is the humanity in HAL that causes the breakdown; not that HAL is human, but because he has humanity *in* him, and the fact that humans do not work logically. That we think we do ultimately leads to great misunderstanding and death.

A Fault in the AE35 Unit[29]

HAL, OF COURSE, IS NOT the only algorithm to cause mayhem (or, more precisely, Dave and Frank are not the only people to put too much trust in automated technologies).

The most recent and most dramatic real-life example was due to a particularly incisive algorithm that ultimately caused the financial collapse of 2009. Like HAL, this algorithm caused massive damage, and, like HAL, it was because it was blindly trusted that massive consequences followed.

This algorithm comes from a family named *copula* (or "coupled"). Copula algorithms come from a need to measure paired, or related, risks. The Gaussian copula algorithm is a statistical approach that's used by banks and Wall Street pros to assess paired risks. So, for example, if there are two horses in a race, let's say Ferdinand and Spend A Buck, the chances that Spend A Buck will break a leg, or fall, or catch the flu are

[29] HAL says to Dave, in *2001: A Space Odyssey.*

all about 1 percent. That's a negligible risk, from a banker's point of view. But if these two horses are running in the same race, and one falls, then that increases the chance that the other horse might also fall. Or if these two horses aren't racing together, but are just in the same stall, and one catches the flu, then there's an increased chance that the other one will also catch the flu.

The error was in trusting the technology.

Anyway, this Gaussian copula algorithm is designed to help bankers clarify, predict, and even design paired risks.

All through the years leading up to the financial collapse of 2009, more and more financial institutions began to adopt the use of this Gaussian copula formula. In fact, it was so successful at predicting risks during the four or five years that led up to the crash, the use of the function was adopted as a means of determining the risks that using the function introduced (the risk was, of course, reduced). If Banker A knew that Banker B was using the formula, they'd establish a kind of trust based on the fact that both of them were using the same method. This actually increased risks, but invisibly, and no one knew it. Not even the inventor of the system.[30]

These financial markets, because they were relying on the same algorithm, started to huddle together. A bunch of different asset-backed securities, such as collateralized debt obligations (or CDOs), were created by mixing together some highly volatile materials, and, as the mixing happened and the stirring sped up, few people seemed to be paying attention to whether the cocktail would explode. Bank loans, corporate bonds, securities that were backed by repeat mortgages, or even lower-rated tranches from other CDOs, were all stirred together in the same big kettle. Some of the most volatile CDOs were composed entirely of subprime mortgage debts. But nobody cared. The algorithm told them it would be okay. After all, it wasn't as if these investment guys were under deep scrutiny. Financial companies are there to make money, and as long as the guy down the line was handing them a sack of cash, they didn't bother to check to see where it was coming from. Why should they? The results were what mattered. Most of the managers of these financial institutions didn't understand how the function worked, anyway.

The algorithm, like HAL 9000, assumed events would cluster around predictable averages. The copula function was used to price CDOs and predict changes, but only within a certain period of time (namely, when house prices were higher). But it's hard to make predictions, especially about the future, and so the unwinding happened as the algorithm began to be influenced by the very factors it was

[30] A quantitative analyst and actuary named David X. Li.

designed to measure. Just like a suspension bridge in a disaster scenario, one side starts to wobble, sending tremors to the other side, which amplifies the effect. Then, combined with the source, it creates a progressively larger wobble. That system wobble, like the wobble between HAL and Dave in *2001*, was based on the same thing: trust in the system.

Many financial organizations trusted the algorithm that created the collapse of the banking market and sparked a global economic meltdown.[31] Was that human error? Yes, of course, but the error wasn't in designing the technology. The error was in trusting the technology.

Our confidence in our technologies can cause us to fall, and far. Especially when the tech is virtual, invisible, or hard to grok.

It is as if, as our technology advances, and we step forward with it, we find ourselves on increasingly thin branches, fingers of technological progress that allow us to take small steps forward, but on narrow walkways that are dangerously weak.

[31] For more on this, see "The Formula that Felled Wall Street," by Sam Jones, *Financial Times*, April 24, 2009.

Chapter Four: Iron Man

How to Dodge Bullets, Leap Small Buildings in a Single Bound, and Stop a Speeding Train—Body, Part 1

> The computer is the most extraordinary of man's technological clothing; it's an extension of our central nervous system.
>
> —Marshall McLuhan

CLOSING MY LAPTOP, I ALSO CLOSED MY EYES AND then put my head back against the seat. I listened to the rhythmic pounding of the old train, *whumpump, whumpump, whumpump,* spent a few minutes of quiet, maybe dozed off; then, I turned to look out the window. Outside were those slightly curved roofs, and some school kids with little square backpacks, and even a few crows flying alongside the train. One of them seemed to look in at me. There were lots and lots of crows in Tokyo.

Pulling the window down, I stuck my face up and tried a bit to get a whiff of the air. It smelled fresh. The wind smelled like rain and the sea. Everything seemed quite real, except for where I was going.

I was on my way to visit a company named Cyberdyne, north of Tokyo in Tsukuba. Tsukuba is one of the world's largest coordinated government efforts to support and improve technical development, a key investment that the Japanese government made to ensure that the tiny island nation would stay ahead of the world's technological curve. Almost half of Japan's public research and development budget is spent in Tsukuba. Modeled in part on Palo Alto, California, in the heart of Silicon Valley, Tsukuba was officially designated a technologically minded city in the 1960s.

By the year 2000, Tsukuba had a population of nearly 200,000 and boasted 60 national research institutes, 2 universities, and nearly 250 private research labs, with hundreds of researchers from around the world visiting these labs and helping to disseminate Japanese technologies back home at any one time.

This is still the case today, but things continue to grow. Since World War II, Japan's government-industry coordination, work ethic, mastery of advanced technology, and very small defense allocation[1] have helped Japan scoot fast to the top of the pile as the second most technologically powerful economy in the world, after the U.S. Japan is the third largest economy in the world after the U.S. and China. More than 90 percent of its citizens are online, and it has more Internet hosts per capita than any country in the world. Much of this is owed to Tsukuba.

Kyoga was wearing an exoskeleton, a "wearable robot."

I opened my laptop and continued reading, where I'd left off, about a cyborg in Switzerland.

In the Alps near Zermatt is a 4,164-meter (13,662-foot) snowcapped mountain named Breithorn. There, on a bright August day in 2006, a young man ascended the mountain, pulling a sled behind him over the snow. He was followed by a team of researchers. The person pulling the sled was Kyoga Ide, a Japanese teen plagued with muscular dystrophy, which slowed him down significantly, even at age sixteen. Underneath his down parka Kyoga was wearing a full-bodied exoskeletal suit that was wired directly to his skin. Small sensors picked up his movements, and, as he flexed a muscle to lift his leg, his exoskeleton assisted him, simultaneously lifting the artificial leg, allowing him to find his way up the mountain and pull the sled behind him, completely on his own.

In the sled rode Seiji Uchida, forty-three, the victim of a 1983 car accident that rendered him paraplegic. Kyoga's exoskeleton allowed him to step over the ice and, via the suit, pull Seiji along with him during the ascent of Breithorn.

The expedition, which had already been postponed once due to bad weather and technical problems, began in the cable-car station of the Matterhorn glacier, 3,883 meters above sea level, and wound its way up the mountain's southwestern flank, over a glacial plateau, before climbing to the summit on a 35-degree snow slope into the high-altitude alpine sun.

Eventually the group had to turn back because the ground became too steep and too icy, but it was an excellent demonstration of two important features of a new technology. First, it showed how our physical handicaps may be overcome, allowing people to participate in physical activities in ways they would otherwise never have been able to do. Second, it showed the world how exoskeletons can aid rescue workers in bringing injured or wounded people from places that might otherwise be too difficult to access—even for those of us lucky enough to have healthy bodies.

Kyoga was wearing a suit designed by Cyberdyne, the Japanese firm that has begun licensing an exoskeleton, a "wearable robot" (as they call it on their Web site[2]), to improve physical

[1] Only 1 percent of GDP (via http://www.cia.gov/).

[2] http://www.cyberdyne.jp/English/robotsuithal/index.html.

Mark Stephen Meadows

Cyberdyne's new offices in Tsukuba.

Mark Stephen Meadows

Inside Cyberdyne's brand new office building.

performance. A wearer attaches sensors to his skin. The sensors pick up what the wearer is doing and relay that information to a "robotic autonomous control system" that drives servomotors in the suit. This allows natural movement that is coordinated with the exoskeleton. In other words, it is an early version of Tony Stark's armor. Short for "Hybrid Assistive Limb," Cyberdyne markets the suit under the acronym "HAL."

Hopping off at my stop in Tsukuba, it was an easy turn down some escalators, through a tollgate that took my ticket, and across the street. The Cyberdyne research office, in the center of Tsukuba, was a glossy-looking office that gave me the impression of stepping into a military complex. It was sleek enough to be housing radioactive isotopes and teleportation devices, and had a cold, corporate look.

Appearances did little to indicate what was inside.

Cyberdyne's HAL

THE OFFICES WERE EMPTY. THERE WAS NO SECRETARY, no bustle, no people slamming phones and

Mark Stephen Meadows

An image of the Cyberdyne exoskeleton at the top of the stairs in Cyberdyne's office.

running across the room. It was simple, austere emptiness.

There was no furniture, either. There was nothing except a carpet.

I called out "Hellooo?!" and it echoed.

It smelled like plastic. It smelled like new.

Soon a small woman named Fumi came to greet me, handed me some documents, brought me upstairs, brought me some tea, and sat me down so I could wait for the CEO of Cyberdyne, Yoshiyuki Sankai. Fumi's presence had been so sudden and so surprising that I had the feeling I was in some postapocalyptic scene where everything had disappeared. But I thanked her, obediently sipped my tea, and began to read the documents she had handed me as she left the room.

Sankai-san has been working on the HAL project for sixteen years. As a boy he dreamed of a suit that could help people be stronger, and his lifelong love of robots is a well-documented story throughout Japan.[3] One of the things that he set out to do with Cyberdyne was to establish a "suitable relationship" with technology, and to make sure that the technology he was building was there to help people and not to be used as a weapon. To that end he first established a working philosophy of "technology supporting people" and went on to build committees and teams that he then spread to municipalities and government-level meetings. He has met with the last four prime ministers of Japan to help establish legal and ethical guidelines for how technology, and, specifically, assistive and prosthetic technologies, should be used.

He first established a working philosophy of "technology supporting people."

Born in the late 1950s in Okayama, Sankai attended the University of Tsukuba. When he saw the number of students who were confined to wheelchairs, stricken with some form of paralysis, usually from an automobile accident, he adjusted his career bearing a bit. He questioned whether standard medicine would provide him the necessary space to make the changes he wanted. After discussions with a friend, he changed his focus to medical technologies and began a careful study of the human nervous system, with a single goal: to create a system that would collect impulses from the nervous system and send them to a machine.

By 1997, Sankai-san had a prototype running. It was rough, but the core of it was there.

As the system improved and he won awards and recognition, opportunities came knocking, and Sankai-san was put in a position to test his philosophies. One of these times was when the U.S. Department of Defense in Washington, D.C., offered to work with him on a robot for military use. He declined, telling them that there were already enough people with missing limbs. He also declined a similar offer from the government of South Korea, sticking with his original philosophy of benefiting humankind, and supporting those in need of help.

As I read my notes I heard a shuffle in the hall. Sankai-san stepped into the room.

He had a big smile, a beige trench coat, and smoked glasses. He removed his jacket, shaking the rain off of it, and I was surprised that he extended his hand to shake mine. Normally a bow would be in order here, but he was obviously used to Westerners and our hand-clasping ways. He wore long hair with a sense of humor and he carried a sparkle in his eye, as if he was about to start laughing.

"We have to consider a future in which the effects of technology have had time to ripple and

[3] http://www.youtube.com/watch?v=iT7IXfcifHE.

impact the world in subtle ways," Sankai said. "For example, look at what the automobile has done. Unfortunately, we don't have enough time to consider our futures. We develop, create, and rush ahead." He shook his head, and between the notes I had read and the passion I was seeing firsthand, it was clear that Sankai-san won't be letting go of this idea until it's pried from his cold, dead fingers.

His vision had pushed him to create a fine piece of technology.

The HAL suit is something between a cane, a car, and a clothing line.

The HAL suit, something between a cane, a car, and a clothing line, is a simple carbon-fiber structure that fits around the outside of the body. It can be resized, reshaped, etc., but the core essence is the same as a bug's: It's designed to support the soft material inside with a chitinous shell. The entire suit is capable of carrying upwards of 150 kilograms (about 331 pounds), and it weighs about one-eighth of its lifting capacity. It has small sensors that attach to the skin and sense the electric signal the brain sends to the muscles. A person that has had a stroke, or has had their legs crushed in an accident, is able to move the suit because although the muscles are not functional, the nerves that connect those muscles with the brain are still sensing the signals. So those signals spark, the HAL suit senses them, the suit moves, and since it is a carapace for the user, the user's limb moves with it. It's as fast as your own muscles.

The HAL suit can be networked with another suit, so if you move your leg (with the sensors attached to the skin), another suit nearby will also move. Or if you and someone else are both wearing HAL suits, and you move your leg, his leg will move at the same time in the same way. Recorded sessions can be captured, and uploads and downloads are possible as part of how the suit learns.

The suit is also a multipart system that can be snapped apart. This means that the bottom half can be used—only the legs—or just one arm could be snapped off and used.

The legs are excellent for helping people walk. For example, there was a man with Parkinson's disease who had been in and out of the hospital for four years. He'd been confined to a wheelchair, and there was nothing that anyone was able to do about it. He was stuck. The patient finally got wind of Cyberdyne's HAL suit, via some rehabilitation workers at the hospital.

Sankai showed me a video of what transpired next. A man in a hospital gown puts on the legs of the suit, with several hospital workers helping him. He leans forward, then stands up. Simple as that. It's as if he were a little weak, but still able to

Photo from NASA

While Cyberdyne works to combine suit and robot to help the disabled, NASA has been working to combine suit and robot to aid in space exploration.

stand up. The man looks around and the people near him start clapping. The next part of the video is of the same man, once again standing, unassisted. Someone steps into the image frame and hands him a walker, one of those square jalopies that old folks use. The guy grabs the thing and lurches forward, one step, then another, showing an incredible kind of hunger for walking (which is not hard to imagine, if you're able to imagine four years in a wheelchair). The third video shows him on the parallel bars, walking between them, again, unassisted, and by the fourth video, he's walking on his own.

"Notice that there is a line that divides cyborgs into two types. That line is the skin."

Now imagine someone whose spine has been severed—the muscles no longer receive signals from the brain because the delivery cable has been sliced. Cyberdyne is developing tools for this as well, in the guise of brain-machine interface (BMI) systems. Quite different from the work that the Rehabilitation Institute of Chicago is doing (which requires surgery on a patient and the physical rerouting of the patient's nerves at the end of whatever is left over of the limb), Cyberdyne is working with a helmet that is becoming very popular here in Japan's research labs.

The system uses two primary technologies: First is electroencephalography (EEG) sensors that measure electrical fluctuation in the head; and second is near-infrared spectroscopy (NIRS), which are sensors that measure blood flow in the head, determining O_2 saturation and the proportion of red blood cells in the blood at any given time. It is totally noninvasive and sits outside the skin. These measurements are then processed and sent to the system, which relays the appropriate signal to the limb, such as "lift."

But these BMIs don't necessarily have to be attached to the HAL suit. For example, they could be connected to another suit that someone else is wearing, or connected to something completely different, like a car. Sankai-san laughs, and explains that he's connected these to hospital rooms, so that a patient can turn off lights, turn up the TV, or sound an alarm, simply by thinking about it.

"A super-cyborg!" I blurted.

He replied, "Notice that there is a line that divides cyborgs into two types. That line is the skin. There are cyborgs that have implants, with technology that has affected them under the skin, and then there are the cyborgs that do not have to have the technology inside them, but instead, they are inside the technology. Most of the systems in science fiction are inside the skin."

"Like Iron Man?" I asked.

"That," he grinned, "was one of my inspirations."[4]

[4] It is worth noting that Sankai conceived, invented, and developed the HAL suit before he even heard of *Iron Man*, which was the movie that came out in 2008. What he was saying was that *Iron Man* has inspired him, not that it was the idea that bore his concept.

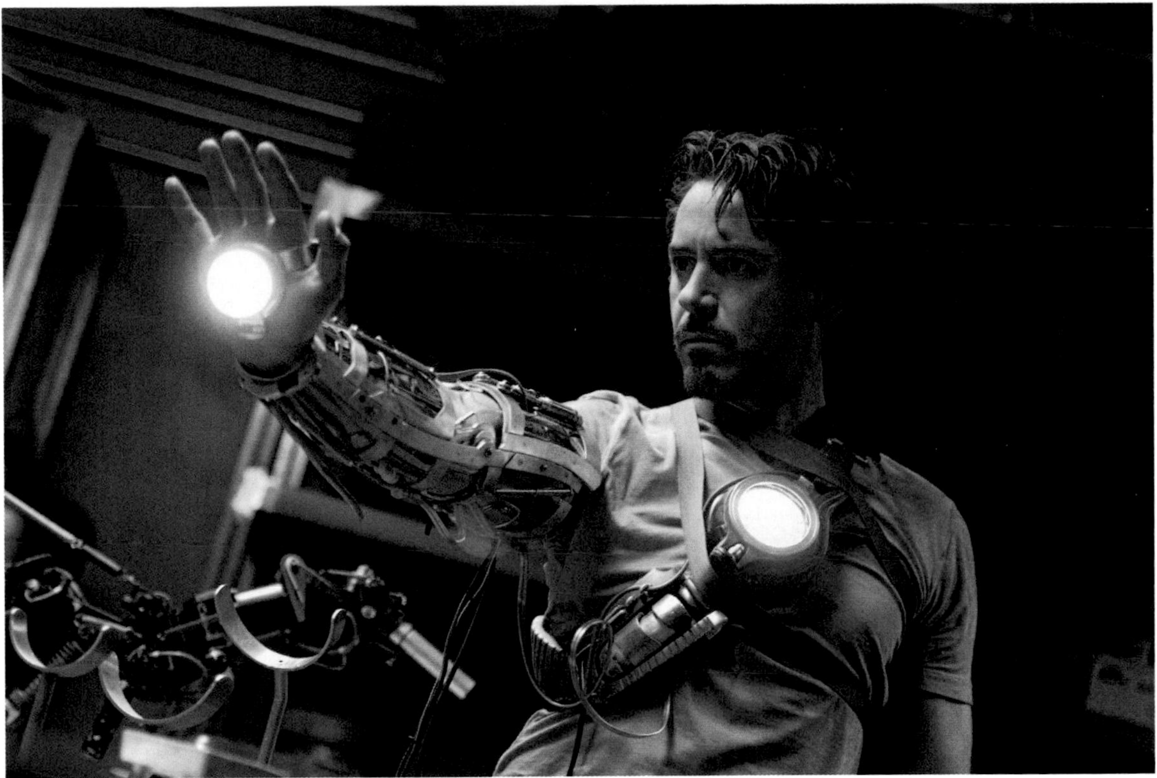

Marvel Enterprises / The Kobal Collection

Tony Stark tests out his new arm.

The Power Suit

THE LINE BETWEEN HUMAN AND ROBOT IS BLURry, at best. If a human is controlling a robot, or a robot is controlling a human, what's the difference between the two? Can people be neurologically connected to machines and drive them around? Can machines be "wired up" to people and drive us around? If a 100-kilogram man is composed of 50 kilograms of machinery that does all the work his body would normally do—walking, digesting, seeing, picking things up, putting things down—and he can control these machines that are attached to what's left of his body via thought, then is he still human? Is the cup half-robot or is the cup half-man? And, more important, when will all—or most—of this become reality?

Iron Man[5] made his first comic-book appearance in 1963's *Tales of Suspense #39*, and when the first *Iron Man* film made it to the big screen in 2008, it generated more than $98 million in Canada and the U.S., in the first week alone. It

[5] *Iron Man* was created in 1963 by Stan Lee, Don Heck, Larry Lieber, and the inimitable Jack Kirby.

grossed over one-third of that during its opening twenty-four hours, making it one of the twelve biggest opening days in movie history.

As he approaches his fiftieth birthday, Iron Man can be found in dozens of action figures, four video games, and hundreds of T-shirts sported by millions of people around the world. This well-loved sci-fi robot shows no signs of aging.

First and foremost, ol' Shellhead's a suit. Though the suit doesn't have a tie, it moves for Tony, makes decisions for him, and strengthens his body. It's a prosthetic body linked directly to his brain. And it's stuck inside of him, too.

First and foremost, ol' Shellhead's a suit.

The movie's story (which very closely follows the original) goes like this: Engineer boy-genius Anthony Stark rises to fame in his prime as a billionaire capitalist playboy. Everything's going great for him. He's a weapons salesman, rich twice over, and drunk half the time. He's untouchable politically, socially, financially, and physically. He does what he wants, when he wants. But things go sour when he's kidnapped in a foreign land by enemy insurgents. Knocked out in a blast, he wakes to discover that he's imprisoned in enemy caves with a new pacemaker installed in his chest, keeping him alive. On top of it, Tony won't be released until he builds his captor a weapon. He has three months.

But being both a genius and a rebel (like any good superhero), Tony surprises us. Instead of building a weapon for his captor, Tony builds an exoskeleton for himself, slugs his way out of his prison compound, and, after a series of life-changing revelations, returns to civilization and his high-tech workshop where he can refocus his heady energies on perfecting another round of new, save-the-world robotic mechanisms. What he builds is a prosthetic exoskeleton that he can control by movement, voice, and thought. After also creating a new pacemaker to go with it, Iron Man is born.

At first glance Iron Man is an exceptionally lucky guy with some seriously kick-ass armor. A suit that not only allows him to fly at Mach 3, safely collide with F-22 Raptors in midflight, fire pulse-cannons from the palm of his hand, and take the occasional hit from a tank's antiaircraft missile, but also to control it all with myoelectric functionality—if he thinks it, the suit does it. The suit also comes with a copilot AI system named JARVIS, who manages the armor's systems with a dry corporate personality while Tony is busy joking it up and battling the bad guys. At first glance Iron Man is a man flying around in a spandex tank—a red-and-gold upgradeable Superman.

But Iron Man is a little more complicated than just a guy in a suit.

Marvel Enterprises / The Kobal Collection

Iron Man out for a test drive.

He's an upgraded human, and his technology is far more than skin-deep. Technically, I understand that the suit is gallium-arsenide-enhanced, bacterium-tiled, self-collapsing, and nuclear-powered, built out of small tiles that accordion into place. He is not, technically, made of iron. And he has this heart condition, and that pacemaker. Although he has a tough outer shell, he is also, at his core, technologically dependent—he's a man who reengineered his own heart.

While imprisoned the first time, Tony Stark collects palladium from available electronics and uses it to build a power system to keep his pacemaker going. In essence, his heart is a small nuclear reactor that's used to drive both his exoskeleton and his real body—a machine that produces a circular magnetic field for confining energized atomic particles (one of the most-researched candidates for producing controlled fusion today[6]). When Tony gets back home, he upgrades this "arc reactor" so that it produces, according to the film, 12 gigawatts of power—enough to power a city of more than a million people for a number of years. But despite its massive wattage, if it is plucked from his chest, Tony Stark drifts toward death. His heart—a piece of technology he invented—is both his power and his fragility, his integrity and his frailty.

[6] For more information, look into Tokamak systems, invented in the 1950s by Soviet physicists Igor Yevgenyevich Tamm and Andrei Sakharov.

Iron Man is a bit of a Tin Man. He's the woodsman from *The Wizard of Oz* who seems to have gained his humanity when he lost his heart. This concept of the heart is important in *Iron Man*, and if the first film is watched closely, lines such as "I know in my heart it is right" and "Proof that Tony Stark has a heart" take on golden-path significance, because at his deepest level, at his heart, Iron Man *is* his technology. Technology integrates not only with his internal pacemaker, but also with his armor, powering it as it powers his own body.

If Tony Stark has an iron heart and an iron skin that are parts of his own body, then what else is an iron man if not a robot? He's robotic at the deepest and the most surface levels. Because his heart is a machine, he seems more a robot with human muscles than he is a man in a metal suit. Like someone that went through a sex-change operation, he's a kind of transgender version of a robot—a man that surgically converted himself, via his prosthetics, into a robot, a bio-bender.

Once upon a time, in a cave far away, just after his pacemaker and artificial heart were installed, Tony Stark dreamed up a list of items to include on his suit so he could bust his way out of the cave and not get shot in the process. It went like this.

IRON MAN'S FUNCTIONALITY REQUIREMENTS LIST:

Bulletproof Exterior
Pneumatically Assisted Exoskeleton
Built-in Tools (miniature saw, machine gun, suction cups, blowtorch)
Air Jets (for short-range jumps)
Electromagnetic Functions (repelling bullets and jamming signals)
Oil Jets (for joint lubrication and slipping up bad guys)
Portable Electric Power Supply

These technologies all existed when Iron Man was himself born in 1963. Later in Iron Man's story, when the inventive Mr. Stark returns to his laboratory, he upgrades his whistles so he can show up with a few more bells. The second version of the suit gets upgraded like so:

Strength Enhancements (for picking up heavy things, like cars)
Sensors (for dodging missiles)
Onboard AI Copilot (JARVIS)
Brain-Computer Interface / Myoelectric Links with Suit

Strength Enhancements, Bullet-Dodging Sensors, Onboard AI Copilot, and BMI Action

SO HOW MUCH OF THIS IS DOABLE WITH THE technologies that exist today?

It doesn't look like we'll see an Iron Man suit in the foreseeable decade, but we will see something quite different, and probably more lethal. Also, more humane.

Much of the new technologies will be designed for military use and neurally disabled individuals. Specifically, we'll see battle gear and replacement eyeballs. These will help to create people who are stronger than normal and able to see as well as an average human.

We'll also see implants designed to overcome Alzheimer's disease and possibly even schizophrenia. We'll see people directly plugged into networks.

In Tokyo, on Halloween day of 2008, Honda unveiled a device that's designed to support body weight, relieve stress on the knees, and increase a person's overall walking strength. It looks a bit like a saddle with legs. The core concept is similar to Cyberdyne's HAL suit.

How much of Iron Man's suit can be built with the technologies that exist today?

Honda's chief engineer on the project, Jun Ashihara, thinks that the folks who will put this to use will be the aging factory workers that do a lot of knee bends, or spend hours standing in lines during the day, or even people waiting on customers, at cash registers, that spend an excessive amount of time on their feet. He compares the device to a bicycle, which is a bit what it looks like. The awkward-looking thing amounts to a seat with two spindly legs that descend to connect with the shoes of the wearer. The system's onboard processor keeps track of what a person is doing naturally, then moves the machine to correspond to those movements.

Four months after Honda's Halloween announcement, on February 26, 2009, at the Association of the U.S. Army (AUSA) Winter Symposium in Fort Lauderdale, Lockheed Martin debuted a form of exoskeleton that helps a soldier tote a 200-pound load up a hill without so much as breathing hard. The exoskeleton is designed to augment a soldier's walking strength and overall endurance. Based on a design from Berkeley Bionics[7] of California, the exoskeleton runs on about 250w, and is called a Human Universal Load Carrier (HULC). On their Web site,[8] Lockheed explains it like this:

> *The HULC is a completely un-tethered, hydraulic-powered anthropomorphic exoskeleton that provides users with the ability to carry loads of up to 200 lbs. for extended periods of time and over all terrains. Its flexible design allows for deep squats, crawls and upper-body lifting. There is no joystick or other control mechanism. The exoskeleton senses what users want to do and where they want to go. It augments their ability, strength and endurance. An onboard micro-computer ensures the exoskeleton moves in concert with the individual.*

It's still not Iron Man's kit. Rather than a nuclear-fusion reactor, the HULC runs on two 4-pound lithium polymer batteries. But,

[7] Formerly Berkeley ExoWorks.
[8] http://www.lockheedmartin.com/products/hulc/index.html.

according to the spec sheet,[9] it can be integrated with armor plating, heating and cooling systems, and other "custom attachments," such as the 55-pound remote-controlled gun mounts that you can order with it. But you can only move at about 3 mph, per battery. So for a price that only Tony Stark wouldn't balk at, you get to be locked into something that lumbers along at 3 mph and weighs 61 pounds, too. Fast humans can reach 25 mph over short distances and cover as much as 9 to 10 miles in an hour. You can't even pick up a car with HULC, but it's technology that's operating today in Afghanistan.

"You can't hump a rucksack at 11,000 feet for fifteen months and not have that have an impact on your body," General Pete Chiarelli, the army's vice chief of staff, told reporters at a press conference in January of 2009.

A team at Raytheon Sarcos, led by Stephen Jacobsen, has also developed an exoskeleton. The wearable robot, called XOS and driven by engineer Rex Jameson, is capable of running, jumping, and even speed-boxing a punching bag. Jameson also does long series of repetitions on a weight machine, pulling down 200 pounds. "He stopped because he got bored," Jacobsen says, "not because he was tired."

"The idea is that if you're holding a 200-pound box, it'll feel like 20 pounds."

"Qualitatively the suit has good mobility," says Jeff Schiffman of the army's Natick Soldier Research, Development and Engineering Center, who has worked on the project for several years. He says the XOS provides a roughly 10:1 gain for a human. "The idea is that if you're holding a 200-pound box, it'll feel like 20 pounds," he says.

When Tony Stark slams his fist against the steel door in the movie *Iron Man,* the militants in the cave cower in fear. Maybe we all should. Because if the suit is being deployed in Afghanistan, it may well represent the future of law enforcement.

In the movie, Iron Man easily dodges a tank's missile. This one's not even in the prototype phase, but there is a patent pending.

Let's say a bad guy shoots at you and you happen to be wearing something IBM calls "Bionic Body Armor." If so, you can—at least, hypothetically—get out of the way before the bullet hits, Matrix-style. In 2009, IBM filed a U.S. patent application on this technology[10] for an automated system that can determine if a bullet is headed your way. If so, the system pulses your muscles with a small shock, causing you, for example, to duck. It was designed especially for high-profile public speakers, and points

[9] http://www.lockheedmartin.com/data/assets/mfc/PC/MFC_HULC_Product_Card.pdf.

[10] U.S. Patent Application #7484451. This patent application was later withdrawn, or else the patent number was changed. No further information is available.

out that a bullet that travels 750 meters (about 2,500 feet) takes about four seconds to arrive. If the armor senses that a bullet is coming toward you, there's time to dodge it. The patent application reads like this:

> *A method of protecting a target from a projectile propelled from a firearm comprises detecting an approaching projectile, continuously monitoring the projectile and transmitting an actual position of the projectile to a controller, computing an estimated projectile trajectory based upon the actual position of the projectile, determining an actual position of a target with a plurality of position sensors and a plurality of attitude sensors, determining whether the estimated projectile trajectory coincides with the actual position of the target, and triggering a plurality of muscle stimulators operably coupled to the controller [. . .] The projectile may be detected in the detecting step by emitting an electromagnetic wave from a projectile detector and receiving the electromagnetic wave after the electromagnetic wave has been reflected back toward the projectile detector by the projectile.*

This is quite close to Tony's bullet-dodging sensors; it even uses electromagnetics to detect projectiles before the target ducks to avoid them.

When one takes the time to investigate the rather extreme vehicle that President Obama drives, known as "The Beast," it seems likely that this technology is on the near horizon. There are no component parts of this technology that can't be implemented today, and there is no reason why this technology couldn't be easily implemented, so we can expect this to exist, if it doesn't already, in the coming decade.

So far, suited up with today's Iron Man technology, we're not picking up cars or stopping speeding trains with an outstretched arm, but we are humping heavy loads as we dodge bullets. Assuming these suits don't interfere with one another (which is probably not the case at this point), or break down (which is definitely not the case these days), we'd be set.

As we progress toward the more fantastic on our Iron Man wish list, we come to the onboard navigation system. For example, in the movie, JARVIS makes astute quantitative judgments, such as pointing out that Tony's flying too high, or that his battery power is too low. Once JARVIS has advised action, Tony can take it or leave it.

Automated decision-making to allow navigation has been done since the early 1900s. For example, autopiloting planes and boats have been around for decades, and self-steering is a problem that's been pretty much solved, even in the more-complicated world of automobiles. While not easy, it's commonly done.

Serious research along these lines started in the 1970s, and by 1993 the European Prometheus Project outlined a copilot architecture based on AI systems functioning in real-time environments. With more than a billion dollars in funding from the European Commission, the forty participants (including Ernst Dickmanns, a leading pioneer in driverless cars) developed an autonomous Mercedes-Benz which managed to speed its way along the Autobahn for 158 unaided kilometers, passing human drivers and arriving at its destination with carefully defined and qualified success. That was in 1995.

Ten years later, the U.S. Defense Advanced Research Projects Agency (DARPA) sponsored their second autonomous vehicle competition. Twenty-three automotive systems started this car-shaped robot race, and five of them successfully completed it, each using radically different methods of perception, navigation, and problem-solving. These car-shaped robots were assisted by humans, engaging in "human-assisted" driving. That is to say, the robots were being helped by people.

But the automotive industry's concentration has been on the opposite—on "driver-assisted" technologies (in which the robot helps a person). This, of course, is different from "autonomous" technologies (in which a car drives itself to and from predetermined destinations). While "driver-assisted" and "human-assisted" have both been done, "autonomous" has yet to be realized largely due to a lack of customer desire for an autonomous car. Major corporations such as Daimler-Chrysler (the company with the world's largest private research budget: $5.8 billion as of 2008) discovered that completely autonomous cars are not in demand—even if people don't like driving, they seem to prefer it to being driven. So today fully autonomous vehicles have taken a backseat to "driver-assisted" technologies. In short, people just aren't ready to give up that kind of control to their automobile.

Completely autonomous cars are not in demand.

But this is not so true when it comes to the military. The military aeronautics industries have always been well ahead of the automotive folks when it comes to autonomous vehicles, and if we take a look at Unmanned Aerial Vehicles (UAVs),[11] we find thousands of vehicles that have crossed the line from "human-assisted" to fully autonomous, able to take off, navigate, and land at a given destination, completely on their own. These systems have been used since the Vietnam War, and they continue to be used in

[11] Also called "Unmanned Aerial Systems."

Afghanistan today. UAVs are a kind of hybrid between a robot and a plane, and they are smaller, faster, more precise, cost less, and, most important, remove a human from the battlefield. According to the *New York Times,*[12] the U.S. Air Force has, as of March 2009, over 5,500 UAVs (compared to the 167 they had in 2001), and the number of missions they fly each month continues to skyrocket.

So while here on the ground things seem to be crawling, up in the sky we can find robotic copilots that help human pilots, robotic pilots that help human copilots, and there are blends appearing everywhere in between. The development of these technologies is creating a funhouse hall of mirrors, full of humans and robots, both of them looking more and more like the other.

Lastly, we have our brain-machine interface, or BMI (Tony's myoelectric links with his suit).

In 2008, in the Marvel comic-book series, Tony Stark modifies his nervous system with a technically laced virus (Extremis), which allows him to both store portions of his armor in his bones as well as control it remotely through myoelectric brain impulses. It's a step in the prosthetic direction, and it is technology we can see emerging now.

Brain-machine interfaces (BMIs)[13] are being used to drive artificial limbs today. Since 2007 Todd Kuiken and colleagues at the Rehabilitation Institute of Chicago have been outfitting amputees with robotic arms that feature motorized shoulders, elbows, wrists, and hands.[14] Patients can then wave their arms, pick up crackers, and even grab a small disk as it rolls across a table. The surgery is rather delicate, as it includes plucking nerves that previously carried signals to the amputated limb to muscles in the chest and upper arm. These nerves are then rerouted so they may be attached, just at the skin layer, to thin wires used to transfer these signals. Since the muscles contract naturally when the patient thinks about moving their arm, those signals are then read by sensors on the prosthetic limb, which are moved via a local electric supply. Patients are also able to perform tasks that require multijoint coordination, such as throwing. It's Luke Skywalker's hand, and it exists now.

According to interviews with Dr. Kuiken, the next step is to add sensory feedback to the system so that patients can get rid of a potentially crushing handshake. This work will of course lead to a refined tactile information sense so that patients feel more and more like they have a real hand. Imagine manual workers who have lost a hand at work, and their insurance company gives them enough support that they can return to work with a little extra strength. This seems to be where we'll be headed in the coming decade.

[12] *March 16, 2009.*

[13] Brain-machine interfaces, or brain-robot interfaces (BMI and BRI, respectively) are also known as BCIs, or brain-computer interfaces, and I am synonymizing these terms quite liberally.

[14] As reported in *The Journal of the American Medical Association*, Monday, February 9, 2009.

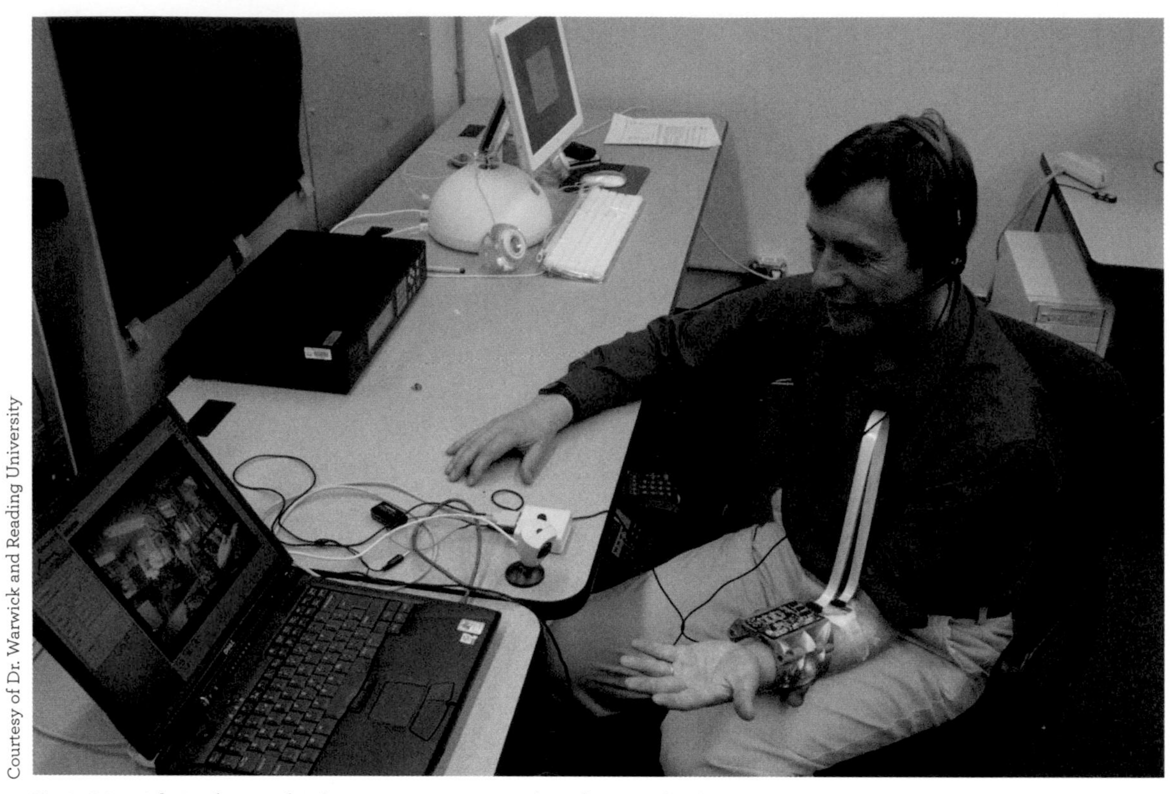

Courtesy of Dr. Warwick and Reading University

Kevin Warwick, Professor of Cybernetics, University of Reading, England.

I spoke on several occasions with Kevin Warwick, a British scientist and professor of cybernetics at the University of Reading, who has not only done a tremendous amount of work toward linking human nervous systems and robotic systems, but also has a neural implant of his own. One of the questions I asked him was where he saw neural implants headed. He told me that there should be no problem when it comes to merely assisting the disabled, but when it comes to human enhancement, who can have the implants, for what reason, and who makes money from it, we'll find some big ethical questions on the horizon.

Embrace Your Inner Cyborg

EXOSKELETONS LIKE CYBERDYNE'S OR HONDA'S help disabled people function in a world that demands physical abilities. We all have to climb stairs, carry suitcases, or bend over from time to time. The goal of these technologies is to help people do just that. In essence, they are strap-on exoskeletons.

Fred Downs is the head of prosthetics for the Veterans Health Administration in Washington,

D.C. Since stepping on a land mine during the Vietnam War, he's been wearing the standard-issue prosthetic hand, a simple hook and clamp device that attaches to his arm. Demonstrating his prosthetic hand's limited abilities for a *60 Minutes* interview, Downs explains, "It's a basic hook. And I can rotate the hook like this and lock it." The technology was developed during the World War II era. "It's something out of *Peter Pan*. And that's just unacceptable."

But in the last four years, DARPA has invested over $100 million in researching new prosthetic limbs for veterans, and, with the help of over 300 engineers and neuroscientists, results are looking more like *Star Wars* than *Saving Private Ryan*. The flagship project, named "Luke" (as in Skywalker), is a prosthetic hand that contains twenty-five circuit boards and ten motors, enough to rotate, pivot, and take commands from small buttons embedded in a shoe that the wearer pushes while operating the arm, as if typing with their feet.

Inventor Dean Kamen (whose past inventions include the Segway and a long list of medical devices) was commissioned to develop a new prosthetic under the project's DARPA funding. The goal was ambitious: to allow a wearer to not only pick up a grape or a raisin, but to be able to tell the difference between them—with his or her eyes closed.

"It felt like my arm. It was me."

One year later, Kamen and his team at DEKA Research and Development had a prototype ready for testing. Kamen asked Fred Downs to take off the hook he'd been wearing for forty years and give the new arm a try. After only ten hours' practice, Downs was able to pick up a soda bottle.

"The feeling is hard to describe," he explained. "For the first time in forty years, my left hand did this," he said as he closed his fist with his other hand. "I almost choke up saying it now. It was just—it was such an amazing feeling. I was twenty-three years old the last time I did that." Downs was clearly moved. "It felt so good to move my arm again—to do things with it. Not as fast, but it worked."

Scott Pelley of *60 Minutes* asked, "You just said 'move my arm again.' Did it feel like your arm, all of the sudden?"

"It did. It did. It felt like my arm. It was me," Downs said.

DEKA and the DARPA-funded team continue their work with new prototypes scheduled for production in the coming years. Their work will surely affect the lives of millions of people.

There are dozens of similar projects happening around the world.

In 2006, Pierpaolo Petruzziello, a twenty-six-year-old Italian man, lost his left

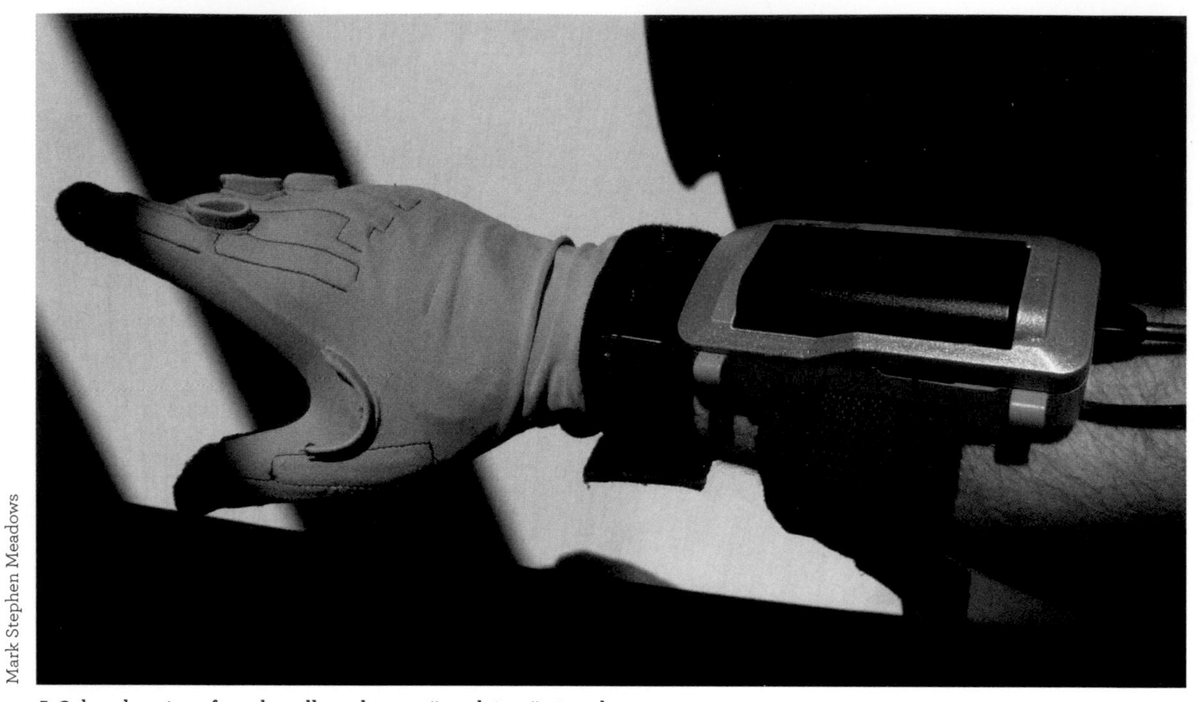

Mark Stephen Meadows

A Cyberglove interface that allowed me to "reach into" virtual space.

hand in a car accident. After working with the LifeHand project, a research project based at the Biomedical University in Rome, Petruzziello now has a thought-controlled prosthesis. In order to install the system, electrodes were surgically implanted directly into Petruzziello's arm, which then sensed when he moved his muscles. Those signals were sent down the line, subsequently moving motors in the mechanical hand. During rehabilitation, after the implant surgery, he had to concentrate very hard to learn to control the hand. He had to imagine that the hand was still there, even though he knew it was not. It was a difficult process that amounted to altering his synaptic structure to control a robotic machine, rather than an organic hand. But Petruzziello was able to control the hand, often from across the room.[15]

There are many cases in which prosthetic hands would be quite handy: dealing with hazardous materials or toxic waste, as well as things that are hot, cold, acidic, electrified, or would otherwise damage a hand; working with sterile or very small items that would require a particularly refined sense of touch—these are all probable applications.

15 . . . and, therefore, anywhere in the world via the Internet.

We can foresee present-day "data gloves" in which molecular engineers or nanotech engineers would be able to directly "grab" and manipulate atomic elements as if they were balls in the air in front of them, thus reengineering even molecules for nanotechnological applications. Biological engineers are already able to directly interface their hand movements with virtual environments that are invisible, interfacing with pure-data environments. Direct tactile manipulation would not only save millions of dollars but also open up entire new industries. And of course, prosthetics and such modifications give the phrase "rapid deployment of a firearm" a whole new meaning.

Soon, prosthetic hands will prove superior to our natural-born hands; prosthetics of all sorts may well become a form of optional surgery, much like cosmetic surgery is today, in which people prefer to have robotic hands to the ones with which they were born.

As I type, I look at my keyboard and see a rather slow and kludgy interface that we could improve upon with prosthetics—perhaps removable gloves that are integrated with BMI technologies. There are thousands of applications that would allow us to better interface with our computers, certainly. We already see people around us wearing Bluetooth headsets that provide them with prosthetically enhanced ears to better interface with mobile computers. Is it hard to imagine a finger that contains a USB drive, which you can simply use whenever you're near a computer?

This was done in the winter of 2009 when Jerry Jalava, a Finnish computer programmer, lost one of his fingers in a motorcycle accident, and made for himself a prosthetic replacement with a USB drive. It has 2GB of memory which he uses to store photos, movies, and code. It's not permanently attached, so he can remove it, but he's already considering upgrading it to include larger storage and wireless connectivity so he doesn't have to take it off each time he uses it.

Meanwhile, we can hear the distant stampede to own the intellectual property that will allow commercialization of these technologies. For example, Intel claims we'll have chips in our brains to control computers by 2020.[16] Microsoft has filed muscle-input patents,[17] and biotechnology companies are also beginning to look in the same direction.

As with all questions of prosthetics, the technology's proximity to the body is the key ingredient. There are two points to be considered: First, where is the input located; and second, where is the output located? The BMI (the input) might be inside the body, or it might be outside. BMIs come in many shapes and forms, from helmets on the head that we can take off to wires in our stomachs that simply cannot be

[16] *ComputerWorld*, "Intel: Chips in Brains Will Control Computers by 2020," by Sharon Gaudin, November 2009.

[17] http://research.microsoft.com/en-us/um/redmond/groups/cue/publications/CHI2008-EMG.pdf.

disconnected. This range of BMIs is the first factor. The range of the prosthetic itself is the second. Is the robotic hand attached to the body, or is it inside of a furnace, in the middle of a volcano, on an island thousands of miles away?

For example, in Petruzziello's case, the BMI is wired directly to his physical nervous system, and the robotic hand is attached to his arm. Both of these are intimately close technologies that are intended to be immediately attached to his body, and under his skin. But a human does not necessarily need to be physically attached to these systems. These prosthetics can take many forms, and they can be networked.

BMIs are not only being used to drive robotic limbs; they are also being used to drive entire robots. Honda Research worked with Advanced Telecommunications Research Institute International (ATR) and Shimadzu Corporation to build a sensor-packed helmet that measures thought patterns and relays them, via a wireless connection, to a robot named ASIMO. If the user imagines lifting his right hand, ASIMO does the same. If the user thinks about moving his or her leg, ASIMO walks, and if the user thinks about moving his or her tongue, the robot lifts his little boxy hand to his little boxy face.

As with Cyberdyne's system, this one also uses the two primary technologies of electroencephalography (EEG) and near-infrared spectroscopy (NIRS). These measurements are then processed and sent to the system, which relays the appropriate signal to the entire robot; for example, move left hand backward, step forward with left foot and now right, now walk, etc.

Of course, this means that we'll not only be able to apply this to UAVs and cars (probably Hondas), but it also means that it will become increasingly difficult to tell if the robots walking around in your neighborhood are being driven by humans or not. What we have here is a kind of physical avatar. Surely these technologies will be used for deep-sea construction, mining operations, and off-planet development in the coming decades.

So, exoskeletons and other systems that put a human inside of the machine are lagging behind the advances of UAVs and other semi-autonomous remote-controlled systems that allow remote control of the system. These systems will continue to improve in the coming decades.

Technologies get faster and smaller each year. Each year cell phones, personal computers, televisions, transmitters, receivers, cameras, recorders, processors, and storage devices all follow this same trend of getting smaller and faster, burning less energy and integrating more easily with other systems and component parts. This is especially true with any technology that relies on semiconductor circuits (such as the helmet for fancy flying red-and-yellow armor). Circuits get faster and cheaper each year. And as the size of transistors has decreased, the cost per transistor has also decreased. Meanwhile, parallel to these developments, we have the fact that networking speeds double every nine months (as per Butter's Law of Photonics[18]), and that wireless reception has shown ongoing improvements in speed as well. (Although opinions vary on how much, it appears roughly equal to Moore's Law, as it is based on the same processing needs.) Parallel to these technologies, other advances continue; propulsion jets get smaller, bulletproof armor gets lighter, etc. The component parts of Iron Man's suit are running now, although it's big, bulky, and in pieces. If we glued these pieces together today, the suit would be a grotesque coupling of dental headgear, alloy-plated motorcycle equipment, and the rather bulky improbability of having a small (and army-green) jet bolted onto your backpack. It would be huge, heavy, and slow. It's just not the sexy suit Stan Lee has promised us. Not yet, anyway.

Most of these technologies exist today; they're just a bit unwieldy. Except for the fist-sized nuclear reactor, we can find that all of the core functions of Iron Man's armor are rapidly becoming smaller and more like something we might—someday, at least two decades out—be able to wear. We're moving in that direction. The core pieces of flight, weaponry, exoskeleton strength, and myoelectrically controlled prosthetic limbs are already functioning on battlefields and in research labs around the world. Even if we can invent fist-sized nuclear reactors that we can wear in our chests to power exoskeletons, pacemakers, and jet boots, I doubt that this technology will ever be used on Earth. The main reason we won't see a suit like this operating on our planet is because robots are already doing most of the work that Iron Man can do—and they're doing it better.

We can throw darts at the calendar and try to pick a year when these technologies will each be small enough to be combined into a suit of armor, and regardless of the year, we'd be off somehow. Should a suit like this appear in the coming decades, it will be secret; the U.S. Army will build it; and it most certainly won't be red and yellow.

While not yet invented, Iron Man's suit is also already obsolete. UAVs have the core functionality of any of Tony's suits, be they the 1963 or 2010 versions. In terms of bombing the bad guys in faraway lands, flying at supersonic speeds, and being fast, small, and hard to hit, UAVs have already out-evolved Iron Man. Predators and Reapers (armed UAVs that are commonly put to flight for military missions) are now flying thirty-four surveillance patrols each day in Afghanistan and Iraq, up from twelve in 2006.[19] These machines—with human/robot interaction—take off from an airstrip and circle war zones with video cameras that stream gigabytes of data to field officers, tacticians,

[18] Namely, that the amount of data optical fiber is able to carry is doubling every nine months.

[19] http://www.nytimes.com/2009/03/17/business/17uav.html?_r=1&hp.

and strategists who can then order an attack. On top of that, it's more or less acceptable if the UAV takes a digger; no pilots will be captured, no lives will be lost, and it costs less than a standard plane. It does not make sense to put a human in a plane—or a red-and-yellow super suit that costs twenty times what a UAV costs—and put his life at risk when you can instead put a joystick in his hand and ask him to sit in a trailer thousands of miles away. The pilot may not be in the machine, but he can still control it as if he were.

Control and autonomy are the weighty topics being balanced as these systems are developed. Autonomy always has some component of unpredictability. Mars Rovers, ocean liner autopilots, or even the simplest cruise-control system in your car are a blend of human and robot, with human control and machine autonomy taking varying ratios of the decision process. All of them, even Tony Stark's high-tech spin on a knight's old-school chain mail, share control at different times. You turn your cruise control off. Tony shuts JARVIS up. So a cyborg is more a dynamic system of sharing power than it is a robot and a human sharing the same body.

The Cyborg Animal

WHAT A PARTY IT WOULD MAKE. IRON MAN, ROBOCop, the Six Million Dollar Man, the Bionic Woman, Inspector Gadget, Pierpaolo Petruzziello, Fred Downs, Luke Skywalker, Darth Vader, Yoshiyuki Sankai, Dr. Octopus, Kevin Warwick, and *Star Trek*'s Borg. Put them all in a boxing ring and let them duke out the definition of the word *cyborg*. The debate's already five decades old.

The term was first used in a NASA research paper authored by Manfred Clynes and Nathan Kline in 1960,[20] referring to the blend of "cybernetic" and "organic" systems for outer space and the astronautics industries:

> *What are some of the devices necessary for creating self-regulating man-machine systems? The self-regulation needs to function without the benefit of consciousness, in order to cooperate with the body's own autonomous homeostatic controls. For the artificially extended homeostatic control system functioning unconsciously, one of us (Manfred Clynes) has coined the term* cyborg.

I don't think they're talking about sunglasses, but they could be.[21] I also don't think they're talking about subdermal myoelectric sensors controlling a robot that's flying over a battlefield a dozen time zones away. But both of these fit into the Clynes/Kline definition.

Then there's the guy at Honda that runs ASIMO via his BMI. He's a cyborg. The soldiers that remotely pilot the UAVs are cyborgs. The

[20] "Cyborgs and Space," in *Astronautics* (September 1960), by Manfred E. Clynes and Nathan S. Kline.

[21] Sunglasses are not technically homeostatic, but they do maintain homeostasis with the rest of the body.

Michelangelo's *Creation of Adam*, c. 1512. Note the controversial Uterine Brain that surrounds God. Some have argued that this is Michelangelo pointing out that God remotely controls each of us. Is this the first BMI?

person wearing a bullet-dodging suit is a cyborg, and so is the person who wears the HULC exoskeleton. They are all people that interoperate with a machine to replace or add to their natural abilities. The person with the prosthetic limb is a cyborg, as is the person in the wheelchair, or with a crutch, or wearing a helmet, hearing aid, or glasses. You don't even need the fancy "driver-assisted" tech to become a cyborg. All you need to do is sit down in your car and turn it on and you, like Tony Stark, meld with your prosthetic armor that helps you move differently. You become Iron Foot, and take on "super walking powers" and transform into a cyborg yourself.

You're a cyborg, too.[22]

We all flicker in and out of our cyborg existence as easily as we turn a key, pick up a remote control, grab a phone, twist a screwdriver, wear glasses, or put on a jacket. It's what we humans do. We use tools to extend ourselves and change our relationship with the world around us, doing it with the mediated devices we make. We need tools to live. Our ability to use tools, build technologies, and create symbolic media has always been dear to us humans, but there's a difference between a caveman wearing a leather jacket and Tony Stark wearing a gallium-arsenide-enhanced, bacterium-tiled, self-

[22] Daniel S. Halacy wrote in *Cyborg: Evolution of the Superman:* "A man with a wooden leg is a cyborg. So is a man with an iron lung. More loosely, a steam shovel operator or an airline pilot is a cyborg. As I type this page I am a cybernetic organism, just as you are when you take pen in hand to sign a check."

collapsing, nuclear-powered, robotic chain-mail suit.[23] Tony's jacket seems, well, unnatural. It's artificial.

What seems to have created the modern legend—the myth—of The Cyborg is the idea that a "natural" human might be welded together with "artificial" technologies. But this is impossible; technology's never been any less natural or more artificial than humans.

The human touch creates the artificial. We create the artificial, just like the legend of King Midas, who turned objects to gold with just a touch. A stone is just a stone until it is used as a scraper. A stick is just a branch until it is used as an arrow. We are the artificial, and we started using tools and technologies the first day our ancestors were human. This idea of "artificial" is a part of being "human."

We've always been cyborgs; our ancestors were cyborgs, and our descendants will be cyborgs. Since part of being a human means using tools, we've just been playing with our legacy of artifacts for the past 50,000 years, or however long it's been since the first human picked up a stick. The first primate that picked up that stick turned it into the first tool just by touching it. Making natural things artificial is one of our defining characteristics.[24]

Today we tend to think of cyborgs in terms of contemporary technologies, as if cyborgs were something new—as if cyborgs were only made out of recently smelted parts from high-tech factories located in the cooler regions of the industrialized world. But that's just not the case; cyborgs are as old as humanity.

Iron Man isn't modern because he's a cyborg. He's modern because he's using digital technologies, because his clothing is autonomous, and because he has technology operating inside the boundaries of his skin. Technologies that are small, hard, networked, and electronic are what make the cyborg named Iron Man historically unique.

We humans, we cyborgs, are shaking off the thick leather and furs as we walk away from the swamps and caves of our past existence. Now, in this electronic age, we walk toward a future that looks more like a factory made of light, and on our path we install devices into and onto our bodies. We click and whir with increasing precision, and we adjust, repair, and improve ourselves with the tools that our parents handed to us. We are translating ourselves into electronic creatures—less like apes, more like information. We no longer wear thick leather and furs. We wear sexy red-and-yellow armor.

It's the natural process of becoming robots.

23 Contrary to appearance, the suit is not, according to the *The Iron Manual* and *Iron Man: The Legend*, made of metal, but rather bacteria that transfer alloys to specific resistance points that are approximately dime-sized. There were many armors over the years, and many designs.

24 Note that this argument applies to other animals, and especially primates (specifically chimpanzees).

Part Two

Futurama

Mark Stephen Meadows

Chapter Five: Blade Runner

On Robots That Will Sing, Search, Make a Mewling Noise, and Then Measure Your Emotions—Emotion, Part 2

> The heart has reasons that reason cannot know.
>
> —Blaise Pascal

ABOUT TWENTY MINUTES NORTH OF TOKYO, I was again on the Tsukuba line, headed toward a famous research center named the National Institute of Advanced Industrial Science and Technology, or AIST.[1]

I had on my headphones and I was reading *Do Androids Dream of Electric Sheep?*, and as I paused to look up I noticed that I was the only one with a book. There was a guy with a newspaper, but the other fifteen people in the car (at least the ones that were reading) each had a small computer. None of them were talking (maybe it wasn't allowed on the train), but they were all feverishly typing messages, evidently sending a text or playing a game or something. They all seemed fervent, frenetic, nearly sweating with the effort of inputting data. They were all pouring some kind of energy into their little computers, whether it was a DS or an iPhone. One young man in particular, who surely had a game, was quite actually hopping in his seat as his thumbs ejaculated a nearly spiritual energy into his keyboard. The woman near him, tapping at her screen with a stylus, struck me as profoundly sad. All I could imagine was that she was breaking up with a lover. The young man with the game impolitely grunted. Across the aisle from them a young girl smiled and her thumbs were a blur as she snapped out phrase after phrase, pausing, hitting SEND, then starting a new missive, each more intense than the previous. There was a businessman, his face over his screen, jaw open, lips slack, and eyes sharply focused on whatever it was he was so engrossed in reading.

All of these people were pouring energy and emotion into the network. Their keystrokes, the files they owned, the tools they used, the pages they viewed, what they said, what they read, and how long they did it for were valuable

---

[1] AIST: http://unit.aist.go.jp/is/main/group/group_e.html.

Ladd Company/Warner Bros / The Kobal Collection

Los Angeles, 2019, *Blade Runner.*

commodities that represented who they each were, what they thought and liked. It was their unique smell, their unique fingerprint, a kind of excretion they each produced. All of them were giving off small bits of personal data; all of them were, without knowing it, excreting something like data-sweat. And as they did so, their machines interactively listened, measured, and collected that data-sweat. What these people were doing was feeding the network. They were giving off energy that the network needs to live. It was as if there were tiny networked bees, collecting the data-sweat that these people excreted.

These emotions are measured and collected because, on the Internet today, there is nothing more valuable. We are the food source for the hive-mind of virtual robots.

"I Believe It's Called Empathy."[2]

IN THE FOGGY CALIFORNIA WINTER OF 1967, ONE of the greatest science fiction authors of all time penned the classic novel, *Do Androids Dream of Electric Sheep?* The author was a strange man who wove stranger realities. His ability to anticipate the future and predict things like fax machines, climate change, and virtual reality has had some readers arguing that he was a visitor from the future. Other readers have argued that he was a clairvoyant, or a writer with a penchant for remote viewing. It's commonly agreed, however, that he was named Philip K. Dick, and it's also commonly agreed that this book (one of many he wrote) was the basis for *Blade Runner,* one of the greatest science fiction movies of all time.

[2] "Garland snapped, 'I think you're right; it would seem we lack a specific talent you humans possess. I believe it's called empathy.' " (*Do Androids Dream of Electric Sheep?* by Philip K. Dick)

Ladd Company/Warner Bros / The Kobal Collection

Voigt-Kampff Apparatus.

As I hope you know, this film takes place in Los Angeles, in the year 2019, and robots are indistinguishable from humans. Genetically engineered androids, or replicants, are manufactured by the Tyrell Corporation, an omnipresent company that got its start creating artificial animals (since by 2019, most animals are endangered or extinct) and eventually came to specialize in androids. The company's slogan reads, "More Human than Human."

Replicants mostly do as they're told. They clean stuff up, move things around, and generally deal with the unpleasant and menial labors humans don't want to do. But this isn't much fun. So, from time to time (as can be expected from a machine with humanlike emotions), a violent outbreak occurs. As a result, replicants got themselves banned from Earth, banished to do their dirty, deadly, or dangerous work on off-world colonies. But this isn't much fun either, and so, from time to time (as can be expected from a machine with humanlike emotions), they give up on all that and return to Earth to get a job, make some money, or seek immortality.

But coming back to Earth isn't legal. So, replicants who return are "retired" by police officers known as Blade Runners.

In both the book and the movie, the protagonist is a Blade Runner named Deckard. Deckard is poor; he's had a rough time in the last couple of years, and he's a mediocre cop. Being a Blade Runner is a dirty deed, and it's done dirt cheap.

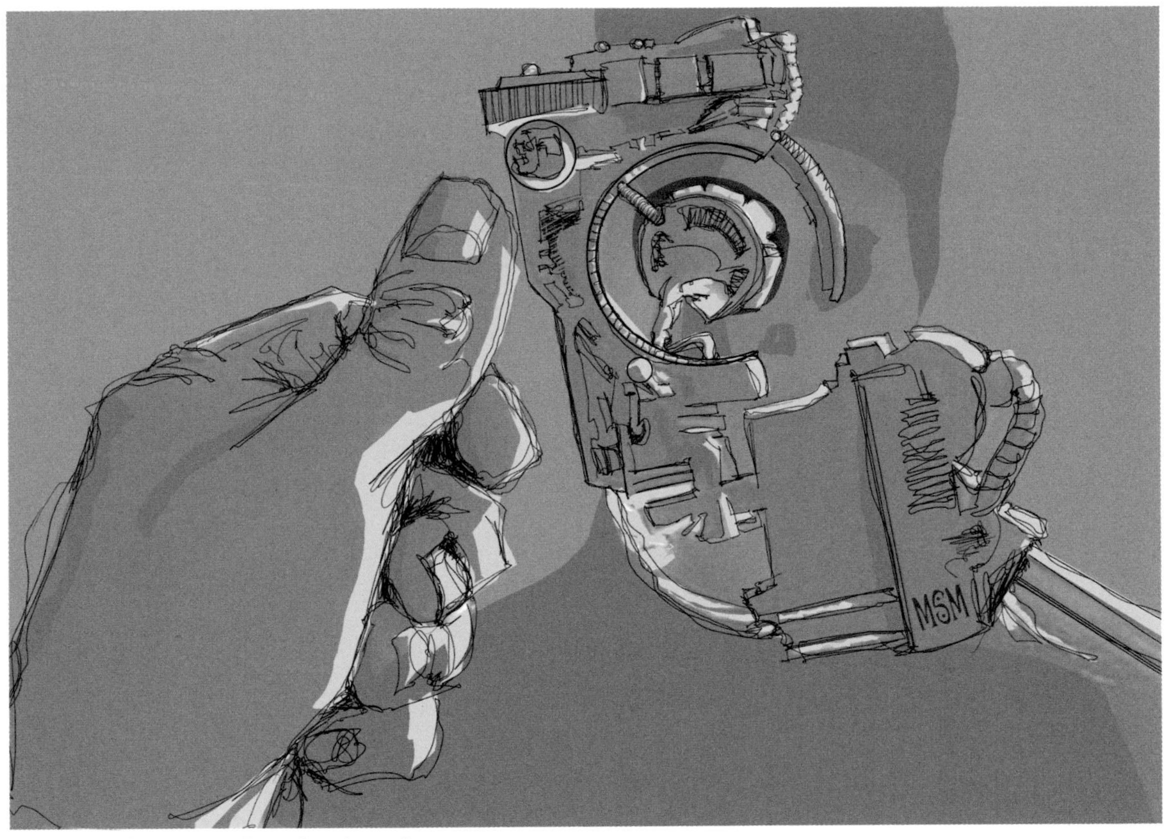

Mark Stephen Meadows

Voigt-Kampff Apparatus from the other side.

But finally he lucks out: Some replicants have come to Earth, and it's up to him to find them.

This is not a simple task. Replicants look, feel, act, smell, sound, bleed, and make love like normal humans. And not only are they stronger than your average mother-born, they're also smarter. So it's hard to catch one, and if you can, you have to spend a little time getting to know it before you can kill it. To make matters more morally complicated, it is not even that the replicant has necessarily done something *wrong*—it's not like they've killed someone and then returned to Earth. They're more like escaped slaves that blend in with the master class, and Deckard has to find them.

Deckard has to measure a replicant's emotions before he can tell that it's not human, since the only major thing that makes them different from humans is the quality of empathy.

To tell the difference between replicant emotions and human emotions, a test called the Voigt-Kampff Altered Scale is used. The

Voigt-Kampff amounts to a "personality profile analytical tool." It's psychological. It measures a replicant's emotional reactions to hypothetical scenarios, and specifically looks for a "flattening effect," evaluating emotions based on pulse, pupil dilation, circulation, respiration, and breathing rhythms. Involuntary stuff. It's a bit like a polygraph, but instead of a lie detector, it's an emotion detector.

The Voigt-Kampff Altered Scale is the opposite of the Turing Test. The Turing Test tells how robots and humans are the same, but the Voigt-Kampff test tells how they are different. In the world of *Blade Runner,* long gone are the days of trying to make robots mimic people. Robots are now so good at mimicking people that tests have been developed to tell them apart.

We are the food source for the hive-mind of virtual robots.

Here in the real world, androids are no longer a thing of science fiction. Androids are real. I've visited with eight and I've had conversations with three. There was no question in my mind about their being robots. I certainly didn't need a test. Their motors whirred, their big camera-eyes clicked, and they had silicone skin that looked like it would be unpleasant to touch. Though they exhibited emotions, I didn't wonder what they were thinking, and I didn't have a hard time telling them apart from humans.

I certainly didn't need to measure their emotions, but behind those clicking camera-eyes, they were measuring mine.

The Stormtrooper Sexpot

WHEN I GOT TO THE AIST I FOUND MY EXPECTED vision of the future. The AIST is surrounded by streets that are as wide as they are smooth, and security guards stationed at each of the entry gates who proudly wear matching hats and suits. The high-tech development complex is monitored by cameras that point from most rooftops, and the campus is shrouded by the calm trees that have been here since this area was a cluster of farms. Little electric cars move quietly between buildings; the only sound that comes from them are the voices of the people talking or the occasional crackling of gravel that gets pinched under the shiny, new tires.

Founded in 2001, the AIST is a combination of more than fifteen research institutes. With more than 3,000 employees, it is Japan's largest nationally funded research organization, and about a fifth of these resources are dedicated to information technology and electronics. Then about a fifth of that group is the subdivision known as the Intelligent Systems Research Institute, and this handful of specialists—some of the most highly experienced researchers and engineers in the field of robotics—have been swimming about in the murky waters of human emotions.

I was sitting on a white leather sofa in an office, situated high above what used to be fields that were plowed by oxen for hundreds of years. Across from me sat Dr. Hiro Hirukawa, the director of Intelligent Systems Research Institute, the group dedicated to robotics research. He had a serious look, a straight mouth and a pressed shirt, and he was very formal and smooth. But there was much more to him than just that first impression. After twenty-three years, Dr. Hirukawa still had a kind of enthusiasm and a sense of humor as he talked, as did everyone in the lab. I was feeling more like I was visiting Zeezo's Magic Castle of Practical Pranksters than the hardened researchers and engineers these guys actually were.

One of the biggest changes in robotics that Dr. Hirukawa has seen during his tenure here has been progress in navigation. Back in the 1980s, there was a great demand for robots to patrol the darkened halls of Tokyo's businesses, late at night, when all the high-flying executives were either out drinking or home in bed. The two challenges, Dr. Hirukawa told me, were to allow the robots to see clearly in these corridors, illuminated only by nightlights, and to move about them appropriately. Walking wasn't the concern because these robots were mounted on wheels, but seeing and navigating were still challenges. These problems are mostly solved these days (robots can now identify facial expressions as well as text on a page), but at the time it was difficult for video imagery to be processed fast enough on a computer small enough to roll around with the Rent-A-GuardBot. The other issue was for the robot to remember where it was. It had to avoid walls, be able to get itself back to a charging station if it was running out of energy, and, as Dr. Hirukawa mentioned, there were the common problems associated with having the robots be robust enough to withstand constant use and the occasional and unintentional abuse of employee handling.

The other big change that Dr. Hirukawa has seen has been in android design. When he started working at AIST, the most complicated robot in the lab was an armature with six joints. Then kids, mostly the kids of the researchers that visited the lab, started to get wind of what a robot was, or was supposed to be, and asked, "Where are the androids?" It was a common expectation, Dr. Hirukawa tells me. A robot had to be humanoid. Soon public calls were coming in to AIST to start this research, and by the mid-1980s it was an assumed charter. It was around this time they began developing the HRP line of androids.

As we continued to talk, we strolled over to the adjacent room where I was introduced to Dr. Kazuhito Yokoi, the HRP project leader and main designer of the HRP system. Dr. Yokoi was a quiet man, with darting eyes and a small curl on the side of his mouth that made me think he, too, was about to crack a joke. The more time I spent with them, the more I realized they were both puppeteers more than scientists.[3]

The three of us stepped out onto a large, smooth floor where the fourth model of the HRP series, 4C, hangs. The robot seemed unconscious, dangling by two ropes attached to its back. The ropes connected to a kind of scaffolding that supported the android, allowing it to be wheeled around. Its head sagged forward, and it looked for all the world like a great, heavy marionette, unplugged. Arms hanging limp at its sides, the hair cascaded, jet-black and pin-straight, down around the face. It was classic Japanese hair. As I stepped closer to the robot and touched the skin, it occurred to me that a robot can have a gender. Just like a movie character or an avatar, this thing was not an *it*, but a *she*. Just as many languages have masculine and feminine words, the worlds of avatars and robots seem to as well. There was no question about the fact that this was modeled with a gender in mind, at least for the head. It looked so lifelike, as if napping, that I noticed I was keeping my voice down, without realizing it, as if I was worried I'd wake the sleeping beauty, startle her, and her head would jerk up.

Courtesy of AIST

HRP-4C before she became so sexy.

Which it did. A small whirring sound rose, the hair bounced back, the eyes opened, the face looked up, the arms flexed, and the feet touched the ground. Small hissing activated HRP-4C's mouth in a manner a little bit akin to an animated blow-up doll. I stepped back to give her room. Her cheeks seemed to puff out a bit as she warmed up. She looked like she was about to spit.

She had great balance for a bipedal machine. Hands on hips and feet together, face

[3] And I mean that in the most complimentary way since puppeteers are artists.

operational, she went into a system test mode. She leaned forward, then she leaned back. She leaned left, then she leaned right. The left arm went up and she bent to the right; right arm went up and she bent to the left. As she rotated her torso I was getting all the signals of watching an athlete prepare for a competition. Then, suddenly, I was reminded of the gears inside and things went uncanny. Maybe from a distance this robot would be attractive on some level, but the quality of the skin was so monochrome (real human skin has many colors under it—even veins are visible) that my brain was starting to reel and refuse what my eyes were seeing. It was repelling. Not that I needed to feel attracted to the robot—it's not like I wanted to have sex with it—but on the other hand, I wondered . . . no . . . maybe I *do* want to have sex with it. Did I?

Why would I have wanted to have sex with a robot? Maybe because it's kind of sexy? After all, for what other reason would a robot be designed as a young, pretty girl if there was not some intention of lubing the gears of ancient urge? Marketing does it all the time, so why can't robotics? Cars can be sexy, and store mannequins too. It seems the robots of the future will get progressively sexier, and that's probably not bad news.

But I was still disgusted by it, or at the very least, found it uncanny to look at.

HRP-4C stands at 158 centimeters (5 feet 2 inches) tall and weighs 43 kilograms (95 pounds). That's with her batteries on. She's got

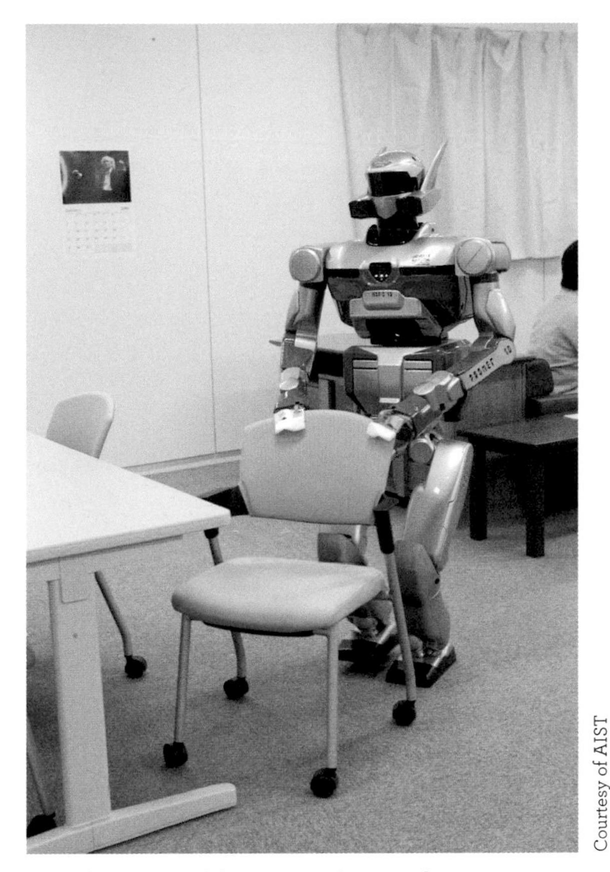
Courtesy of AIST

An earlier HRP model, HRP-2, pushes in a chair.

a kind of stormtrooper body suit on, and just out of curiosity, I asked Dr. Yakoi what her dimensions were. He smiled and said it was a secret.

Looking back at the robot, still warming up, I decided I wanted to have sex with her. This was horribly embarrassing, but why not? I've been turned on by ink on a page before, or pixels on a screen, so why not a robot? But I couldn't exactly figure out why I wanted to have sex with

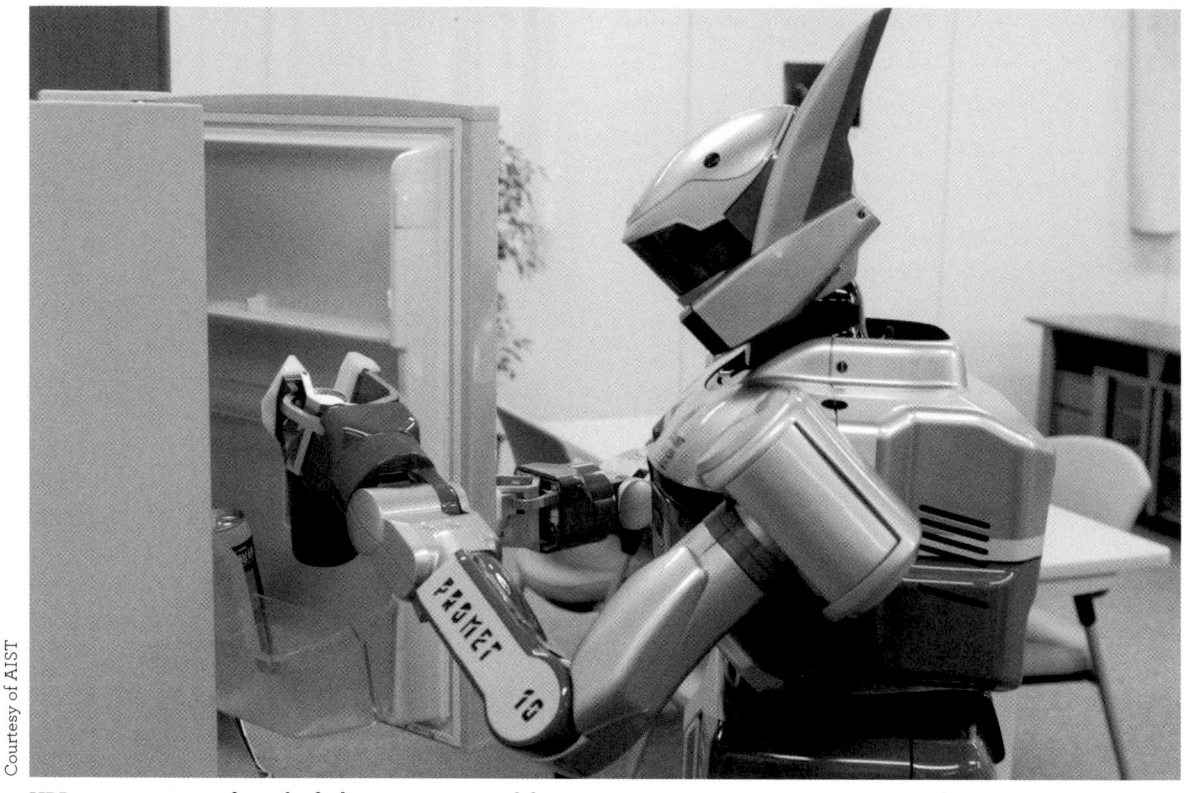

Courtesy of AIST

HRP-2 removes a can from the fridge via pre-scripted directions.

the robot, other than perhaps because of her hair. She had what appeared to be healthy hair. The attraction to my first robot love remained uncanny, and I said nothing of this to the gentlemen that were near me.

HRP-4C continued to warm up, and I couldn't keep my mouth shut, so I asked Dr. Yakoi if he found the robot sexy. He replied, sagely, "She is cute, but not sexy for me because I am her father," and as if to really make my brain shrivel up the sides of my skull, he showed me a video of HRP-4C in a wedding dress.

After some press releases and public shows in March of 2009, it was July 22 of that same year when HRP-4C made her second public debut, and all the cameras were there to bear testimony. Though she'd been seen singing at an electronics conference, and though she was already known as something of a celebrity, this occasion was different.

It was the eighth annual Japan Fashion Week, and Yumi Katsura was showing her Paris Grand Collection of dresses. The crown jewel of the night was, as you can expect, worn by HRP-4C. To the strains of Michael Jackson's song, "Billie Jean," HRP-4C sauntered out onto the catwalk wearing a wedding dress. She announced, "I've put on a wedding dress for the first time. I'm very happy today to wear this dress by Yumi Katsura." Just as with any designer, for Katsura, having her work worn by a robot was the equivalent to having her work worn by a celebrity. The $2 million robot strolled smoothly down the 10-meter-long catwalk without tripping over her rather complex gown, and she even struck a few poses for the cameras as she bathed in the paparazzi sparkle.

Past HRP models.

Courtesy of AIST

Under the bright glow of the lights, all appeared fashionable and cool. But underneath the dress HRP-4C was on the brink of overheating. While the AIST engineers had anticipated this, and modified some parts of the costume to cool things down a bit, they knew that things were getting hot. Dr. Yakoi, in particular, was alert to this, and as HRP-4C turned, he knew that the steps were programmed to land in the right places, but the dress could not get underfoot or the robot would fall, stall, or overheat to a critical level.

HRP-4C stood at the side of the runway with Katsura. The android gave sidelong glances at the audience members of the show. She looked cool, calm, and collected. Then, with the AIST engineers biting their fingernails, the android slowly turned, took a step forward, and successfully walked back up the catwalk. The show came off without a hitch.[4]

After explaining HRP-4C's public performance to me, Dr. Yakoi smiled, put the photograph back on his desk, and stood patiently by as three assistants moved the robot into place and ran some controls from a nearby computer. I walked over to peer over the shoulder of the engineer, and on his screen was a 3-D wire-frame model of the HRP-4C. As the 3-D model raised an arm, the android did the same. Around the 3-D image were various interface controls and sliders, switches and buttons—the strings for this complicated electronic marionette, or *karakuri ningyo*.

Roughly translated from Japanese a *karakuri* is a gag mechanism (gag as in "whoopee cushion" rather than "choke response"). A *karakuri* is a device that surprises and delights for the purposes of entertainment or performance, and sometimes for work. Like any contemporary magic trick that you might perform in public, the mechanism has to be kept secret. *Karakuri* were used in theater, to entertain guests and

[4] Part of the walk can be found on YouTube by searching for "HRP-4C—the robotic bride."

Courtesy of AIST

HRP-4C.

family members, and in rituals or festivals, like funerals or parties. So *karakuri* is the mechanism, while the word *ningyo* means a kind of puppet. This is a conjunction of two words that mean human-shaped. Anthropomorphic. So a *karakuri ningyo* is a human-shaped gag mechanism. An android. An ancient, primitive, Neanderthal version of what we think of today as an android, but a grandpappy android, nonetheless.[5]

When I mentioned this to Dr. Hirukawa, he agreed before gently redirecting me to stand in front of HRP-4C and not look at the computer's control interface. (I realized I'd been looking up the proverbial skirt.)

There was a deep pop, then an ambient hum of speakers; HRP-4C was done warming up, and the servomotors drove the head from side to side as the hair swung to follow the head movements. The movements jerked just enough to be noticeable. There was a kind of acceleration and subtle deceleration at the beginning and end of a movement that a servomotor has a very hard time mimicking. It's harder to mimic your neck muscles with servomotors than it is to mimic a servomotor with your neck muscles.

Other incongruent details emerged: The arms did not move. There was no breathing movement. The uncanny valley of appearance and movement was again blossoming below me, giving me a sense of vertigo as I looked at the impressive technology that AIST had created.

Then, abruptly, she opened her mouth again. There was no flexing of lips or loosening of jaw before it happened. Her jaw just snapped open.

The sound of a young woman's voice came from all around, slightly distorted, slightly plastic, very digital. The subtleties of analog were missing, but she sounded just fine, and having never heard an android sing before, I was treated to an otherworldly experience, as if a diva from a parallel dimension had stepped from her palace-factory for a private rehearsal, dressed in lovely hair, but with a head painted by Edvard Munch.

Leaning in to look a little closer at the mouth that seemed to open into a screaming position (with no sound emerging from it), I noticed how

[5] See www.karakuri.info for more great information on this stuff.

the silicone was bending and stretching when it opened. Her head jerked back, like a horse. Mine did too, as I was a little startled, and didn't want one of those 50kg arms to crack my skull. But it didn't, and she kept singing as I tried not to fidget, despite breaking into a clammy sweat as I stared at a hybrid combination of Jung Ryu Won and the Energizer Bunny. All the benefits of a runway model/superstar singer, but with none of the vanity, bulimia, or cocaine habits.

She sang in English, and the little whisper of the servos moved her jaw.

HRP-4C uses Vocaloid, a singing synthesizer application that Yamaha developed, which is basically a type-and-sing application. So if I wanted HRP-4C to sing a Johnny Cash song, I would type in the lyrics and music notes, and out would come the song. Mostly. There are libraries of singers' recordings that map the vocal qualities of original voices to reproduce the vocals. And vibrato and pitch bends can be added in, creating songs in either Japanese or English.

HRP-4C was developed with a number of tools, most of which came from the ISRI division of ATR, in Osaka. She was developed with the User Centered Robot Open Architecture utilizing Linux, OpenHRP3, speech recognition, and bipedal walking technologies that were previously developed at AIST. The android costs over a million dollars.

A diva from a parallel dimension had stepped from her palace-factory for a private rehearsal.

As the song continued, I started getting fidgety. I was having a hard time simply listening and being quiet. But even more than that, I was having a hard time figuring out why I was rapidly becoming depressed. The robot appeared so diligent. The engineers appeared so diligent. I'm sure I appeared diligent, too, but this was having an odd impact on me and making me feel sad.

Somewhere in my head a set of neurons was telling me that this was a person—a singing young woman, perhaps a sexy one—and yet there was no interaction with the eyes, no acknowledgment of my presence by the "person" singing, no interaction where it would normally exist. There was something so poignant and sad about this song, and the way it was sung; it was as if I was meeting someone that had had their shadow surgically removed, someone that had suffered an accident of a spiritual sort, and was in a pain I couldn't understand.

This singing eased me down the slope of a new Uncanny Valley that worked just like the visual one. Maybe it was cued by the visual uncanniness, or the auditory tonalities, or the fact that I was jetlagged in a foreign land; in any case, I was descending into a new Uncanny Valley. After all, we have at least five senses. We also have a sense of time, a sense of balance, and emotional senses, too. We have social and sexual senses. So I'm guessing that there are many kinds of Uncanny Valleys to go with each of our myriad senses, many new experiences robots will introduce which are just beyond the borderlands of what we humans find pleasant or acceptable or just plain normal. This new medium of robots means that we're about to venture into a land where we lose the map for the territory of emotions.

The engineers were kind enough to have taken the time to run this system for me—they were incredibly nice people and had been incredibly generous with their time—but I was having a hard time looking HRP-4C in the eye. Having a girl robot sing to me, with its putty mouth yawning and stretching in almost no relation whatsoever to the rhythm of this synthetic music, is not something I normally experience. I didn't know how to ask for a break, especially with these men who had worked so hard to get this going. She embarrassed me. I felt like I was at a formal cocktail party and someone had told their somewhat sexy retarded daughter to sing me an a cappella version of AC/DC's "Highway to Hell."

The Form and Function of Android Design (or, Android as Entertainment Media)

THE WORLD'S FIRST ANDROID WAS INVENTED here at AIST, in 1986.

At that time AIST recognized they were on the brink of something important, and they decided to build partnerships to help advance the project. The scope of the work was expanded to include Honda and Kawada Industries, and together, the three of them began developing the HRP series of robots. These androids were first built to test different features, to determine what worked and didn't, and second, to determine what an android was good for.

What could be a worse design for a robot than an android?

The engineers at AIST knew they wanted to build an android, but they weren't sure why. It's like deciding to navigate west, but not knowing where you're going. Non-humanoid robots (armatures, rolling carts, automated cameras, etc.) were functioning quite profitably in factories and warehouses throughout Japan. They were bolting bolts onto cars, guarding hallways in buildings, stacking boxes in storage centers, and so forth. But everyone expected androids.

It was a solution waiting for a problem.

Dr. Hirukawa shrugged his shoulders and said, "Our question was, 'What good is a humanoid?'" I blinked at him. This had never seriously dawned on me. In all science fiction, it is the android that drives the cab, waits tables, steers the starship, saves the princess—and it is certainly the android that murders everybody. But then, as I thought about it, what could be a worse design for a robot? Put the center of gravity up high on a system that is barely able

Courtesy of AIST

HRP-3P takes a shower.

to balance, give it two little pegs to stumble about on, then attach a gripping system which, because it can't be retracted into the body, has to be countered against whenever asymmetric five-tentacled pincers are used. Oh, and give it a thing called a "head" which has no function at all, since the sensory apparatus can just as easily be put in the feet or stomach. What does a robot need a head for, anyway? Or a face?

This is a strange design problem to have because the function is following the form. In the classic parlance of American architects, "The form ever follows the function."[6] It is a given in all schools of design, whether industrial, architectural, or graphic, that a good design will serve the function before the form is developed. Here the form was established as a kind of *a priori* necessity, and only then was the function to be determined. And that function was to be determined based on the form. It is a bit like deciding that an airplane needs to be shaped like a big-eared elephant because that is what consumers want. This introduces a huge range of problems, of course, because

[6] American architect Louis Sullivan's 1896 article, "The Tall Office Building Artistically Considered."

big-eared elephants were no more designed to be airplanes than humans were designed to be robots.

As it turns out, an android's form is an android's function.

The question of "How does a humanoid robot walk?" has become "How does a humanoid robot avoid falling?" Right now, as with the varied field of machine intelligence, most human three-year-olds can manage more than most androids. Getting an android to walk is no small task.

An android's form is an android's function.

During the early '90s, this trio of AIST/Honda/Kawada built the first model of the HRP series, the HRP-1. This robot had the ability to see and process incoming data and had some navigational autonomy in its onboard systems. But it had very little navigation ability. In other words, it could see and make its way around a room, but it couldn't walk. It was unbalanced. It had stereo vision, but would get knocked over if so much as a window were left open. By the time the HRP-2 was in operation, the companies each split apart to chase their own particular ideas about how to best develop androids. AIST then grew its capacities for a more-independent research wing, which eventually became Dr. Hirukawa's department. The HRP-3 system was the result of those efforts—a waterproof, dustproof android with over two dozen articulated joints, each serving multiple functions for supporting itself, or for manipulating objects. Having a robot walk on two feet makes little to no sense when you can just put the thing on four wheels and snap on a pair of hands, like the SWORDS robots in war zones. But then, of course, it's not an android. So walking in an uneven landscape, or dealing with even the simplest of stairs, is still a real problem today.

After multiple iterations of the HRP models, and despite progress in walking, the ISRI group was still having a hard time deciding what to do with the thing. Why do we need walking androids? To hand out tickets? To greet people at reception desks? To serve food? Some of those tasks required walking, some required a humanlike form, but a walking, human-shaped robot? What's an android good for?

Dr. Hirukawa looked at me and raised an eyebrow, put one finger in the air, then slowly said, "Entertainment."

Looking back at the robot, which I was embarrassed to still find sexy and very disturbing, then looking back at the good doctor, I wondered what he meant by *entertainment*.

Entertainment is emotional. When we go to experience something like a movie, and we call it "entertaining," that's usually because it hits a broad array of our emotions. The more entertaining it is, the more emotions we feel. The best-selling entertainment media authors of all time, be they James Cameron or Alfred Hitchcock, Mozart or Led Zeppelin, press a wide range of our emotional buttons.

That emotional entertainment almost always involves another person. When we go to the theater, or listen to music, or read a book, it almost always involves a story about someone. Someone we're supposed to identify with. The emotions they feel are the emotions we feel. The reason we identify with a character is partly because of a particular set of neurons called *mirror neurons*. Some of our reactions have to do with releasing various chemicals, such as vasopressin or oxytocin, chemicals that are released when particular media that is believable enough confuses our brain, tricking it into understanding the media as reality. These, and other things, create an emotional response. This applies to all media that presents a person or something that only vaguely resembles a human. That media may be robotic. An android can fill this bill. We can identify with them and therefore find them emotionally engaging and entertaining.

So, evidently, the android is a kind of emotion machine. At least, that is, if they're entertaining. When we think of walking humanoid robots doing things with us, or to us, we need to understand that there is a big emotional component to this. We only experience the Uncanny Valley in androids, and only when the robot offers some prospect of emotional engagement.

As emotion machines, androids offer a potent form of entertainment medium potentially stronger than text, movies, or interactive narratives. Where else is the possibility of dialogue more likely than with an android? Where else is the possibility of companionship more likely than with an android? Where else is the possibility of sex more likely than with an android? Where else can we see the promise of robots more clearly than in the android form?

Designing an android becomes a problem, especially when we consider scale. The bigger the android, the bigger the consequences—not only technically and financially, but emotionally, too.

First, technically: If you design an android with a larger body, it can do more (like open doors), but this also means that it's heavier, and will require more motors and sensors and plastic gears to keep the hips and knees and neck and arms rotating. This means it needs more battery power (and batteries are heavy). This means that human-size androids weigh about as much as a human-size person. And since they're made of hard materials, you don't want one falling on your foot.

Life-size androids are also really expensive. Manufacturing a couple square meters of skin-colored silicone, embedding it with sexy black pin-straight hair, and then stretching that around a plastic skull will cost you an arm and a leg. Then we have the batteries and usually thousands of moving parts, all of which are custom-made these days. So the larger the android, the higher the manufacturing cost.

But the biggest problem with being human-size is the human emotions.

The bigger the android, the bigger the emotion. A larger android represents, on some instinctual level, a physical threat. This is part of why a human will experience the Uncanny Valley with a human-size android more than with a Barbie-size android. It's like this with most toys. Give a kid a toy dump truck the size of his hand, and no one's too worried about it. He'll quietly purr in the corner while the adults sit around the dinner table. No one's going to be too worried until he makes a gagging noise choking on one of the wheels. But if we give that same kid a remote-controlled 2002 Peterbilt, fully loaded with a 380hp Caterpillar diesel engine, air ride suspension, and an 18,000-pound steering axle, none of the adults will be sipping Chardonnay. Or give a little girl a Barbie and it's all adorable and wholesome, but you will likely stop to scratch your head if you happen to discover a life-size Barbie in an adult's bed.[7]

The bigger the android, the bigger the emotion.

So the android's size is proportional to the emotions it stirs, often a threat-based emotion. When I've seen little androids waddling around on the floor of a museum or a friend's apartment, I think they're cute. They are little, and I often pick up little things. When I see the Terminator, I do not think it is cute. I think it is big, and it can pick up big things, including guns, and me.

Philip K. Dick proposed that it would become difficult to differentiate between androids and humans. That's already happening, and will continue to happen in the coming decade. The reason is because androids are an entertainment medium with which we can identify, and, as they become more like us, they will start to identify with us, as well. This emotional interaction will flow in both directions.

More Animal Than Animal

IN *BLADE RUNNER*, DECKARD TRACKS DOWN A RATHER sexy singer android named Zhora. She's working in a concert hall, carrying around a big fake snake, and when Deckard gets to her dressing room to question her and try to determine if she's a replicant, it takes her even fewer questions to determine he's a cop. The interview goes both ways, and the replicant determines that the man is a cop before the cop determines that the woman is a replicant.

In another case of science fiction prescience, robots will see who we are before we see who

[7] All over the world we can find hundreds of thousands of tiny androids that can walk, almost all of them being used as toys. There are literally thousands of walking robots for well under $100. Produced by Sony, Lego, and other large toy companies, these are an incredibly important part of the robotics industry, but I have had to refrain from featuring them here.

Courtesy of AIST

Paro.

they are. They will know our emotions and physical sensations often before we do.

One of AIST's most successful robots—and one of the most emotionally engaging robots to date—is about the size of a newborn baby (it's not an android). The robot's name is Paro.[8]

Paro is a small seal-shaped robot with whiskers that, when touched, cause the robot to make a slightly sick mewling sound and turn its head in the direction of the touch (usually). It can do other, stranger things. It will lift its tail if it is patted on the head, a bit like a cat does, and it will make different mewling noises if you cover its eyes or poke it, hard, with your finger. When I first met Paro, I tried not to think of it as a game controller with buttons hidden under a synthetic fur coat. I tried not to think of it as a Furby.

The little robot is modeled after baby harp seals (*Phoca groenlandica*) from Canada's

[8] http://paro.jp/.

Courtesy of AIST

Paro in the old folks' home.

northeastern coast. The researchers from AIST went there and developed behavioral models based on the baby seals they found off Madeline Island. The resulting robot uses a 32-bit, RISC chip CPU, stereoscopic optical sensors, surface and whisker sensors, a temperature sensor that adjusts its body temperature, and a microphone for voice recognition. If you spend time with the robot, you get feedback from eyelids, paws, and tail, and those sounds it makes. It has a daily and nightly behavioral rhythm, has a kind of mood that's based on how it is being treated. It can recognize its name (given by the user), and makes a lot of mewling noises, especially when being fondled. The thing will, in fact, make mewling noises for an hour and a half before it needs recharging. Oh, and it is covered in antibacterial fur.

Paro was originally designed by AIST to be entertaining, but with such big, helpless eyes and such small, strange noises, AIST noticed that people became increasingly attached to it. Especially old people. The Japanese newspapers *Chunichi* and *Jiji* both reported that research showed Paro was "soothing to the nerves" and "therapeutic," and this launched a

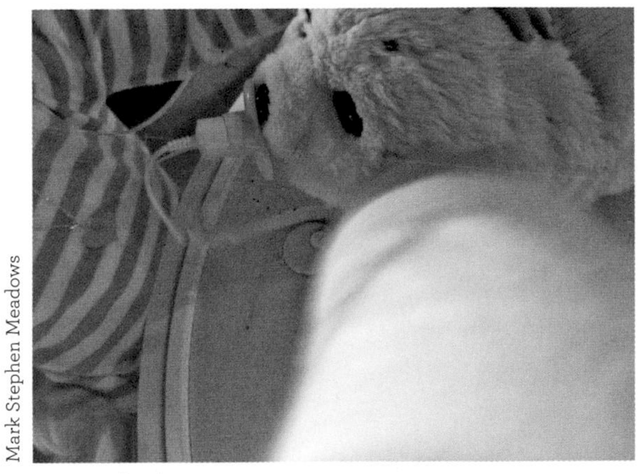
Mark Stephen Meadows

Paro on display.

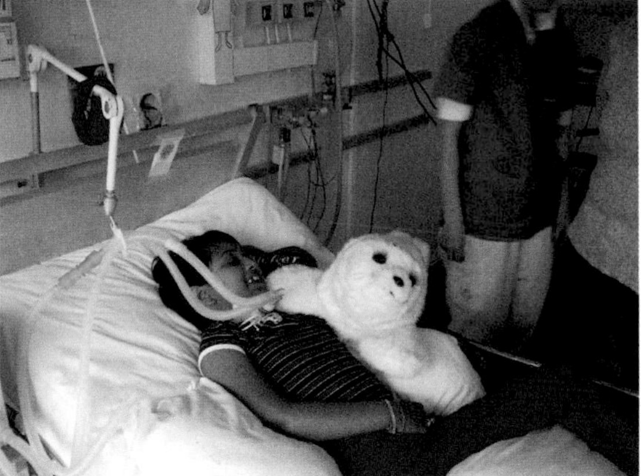
Courtesy of AIST

Paro in the hospital.

small avalanche of media until, eventually, little Paro made it into the *Guinness Book of World Records* under the title "World's Most Therapeutic Robot."

The research that AIST carried out covered psychological, physiological, and social aspects of people's daily life. ISRI/AIST took photos of people's faces as they handled the robot, they interviewed nurses, they recorded what the subjects had to say, and they measured the levels of toxicity in the blood and urine, among other tests. The conclusion was that while handling the robot, people generally were put into a good mood.

As I handled a version of Paro—a solid white model—I could see why people grew attached to them. It moved gently from side to side and squirmed a bit, sensually and slowly, kind of like a puppy. This made me want to protect it a bit, although it seemed independent enough not to be needed, which made for an emotional reading of "cute."

Dr. Hirukawa explained: "It is probably impossible to publish a paper proving that Paro is curing people." He laughed and shrugged, adding, "But if he made it into *Guinness*, well, great. They're not scientific, so that's fine with us. But we've made thousands of these, in many different colors, so something is working." This just made me think of a field full of little Paros, and someone with a baseball bat and an interest in antibacterial fur.

I looked at Paro and Paro looked at me and mewed.

This robot is found in many homes, and now largely lives in senior-care and welfare institutions in more than twenty countries around the globe. Denmark has imported thousands, and

Mark Stephen Meadows

Paro on display.

I've heard rumors from anonymous sources that claim several ministers are proposing a free Paro to each citizen of Denmark.

Spending all this time with the elderly means that Paro has some responsibilities, too. The system can monitor the health of a user, and if the robot is not touched, or is thrown, or gets wet, or hears something that doesn't seem right, that information can be sent back to a central server which then sets off an alarm that can be relayed to mobile devices, as well as other emergency alert systems.

Though nurses and doctors commonly spend time with these elderly patients, Paro can serve as a backup caregiver, which, in theory, allows the nurses to spend more time with other patients if there is a shortage of help around the nursing home. What is far more likely, however, is that Paro won't allow nurses to spend more time with each patient, but less time with more patients.

Another possible solution, and problem, is dementia. Many dementia patients find great comfort in a teddy bear, and if you drop one in their lap, they'll quietly pick it up and play with it. But this might not work as well if we drop Paro into the lap of a dementia patient with, say, an aversion to moth larvae.

How Robots Measure Your Emotions Via Your Body

ON THE SUPER BOWL SUNDAY OF FEBRUARY 1, 2009, more than 100 million Americans were watching the Pittsburgh Steelers warm up, poised to bump heads with the Arizona Cardinals. And in a conference room in New York City, thirty-nine people sat around a television, tossing back burgers and beer, and occasionally adjusting the clumsy black vests they were wearing while watching the game.

The system built into these vests was developed by a market research company called Innerscope Research, from Boston, Massachusetts. They use the system to monitor television viewers as they watch advertisements. Since advertisers are paying nearly $3 million for a thirty-second advertisement, it is in their best interest to make sure that their ads are sent over the airwaves, through the televisions, into the living rooms, and into the minds of their viewers, creating the appropriate emotional response when the message finally hits the appointed target. The monitoring equipment uses three EKG pads and stickers that send data to a computer, which builds an emotional analysis based on the input.

The test amounts to a "personality profile analytical tool." It's psychological. It measures a person's emotional reactions to hypothetical scenarios and specifically monitors pulse, muscle dilation, circulation, respiration, body temperature, and breathing rhythms. Involuntary stuff. It's a bit like a polygraph, but instead of a lie detector, it's an emotion detector. We might call it a kind of Voigt-Kampff Altered Scale test.

Aside from text and pulse, another way to measure emotions is through voice. Several companies have been using "speech analytics" software that monitors conversations between CRM agents and customers on the phone. One supplier of this kind of thing is an Israeli company named NICE Systems. Their house specialty is emotion-sensitive software and call-monitoring systems for companies and security organizations. Their Web site says they have more than 24,000 customers worldwide, including the New York Police Department and Vodafone.

Then there's the face. Researchers at Nanyang Technological University,[9] food and consumer goods company Unilever, and many other companies have produced facial recognition software that takes the points of the mouth and eyebrows, most simply, and builds a map that can be used to create a kind of classification chart. This is then compared against

[9] Dr. Cho Siu Yeung, Nguwi Yok Yen, and Teoh Teik Toe.

known emotions and, just as you and I recognize that a smile most likely indicates pleasure, the software makes a guess as to what the subject is experiencing emotionally.

Smiles are hard to measure, but we're learning more about how to do it. Take, for example, a 2004 study of eighty-one adults at the University of Pittsburgh in Pennsylvania.[10] Using comedy videos, this research compared forced smiles with spontaneous smiles, and showed that spontaneous smiles are more complex, with multiple rises of the corners of the mouth.

Another way to measure emotions via the face is to measure the quantity of blood that gets pushed around. Ioannis Pavlidis at the University of Houston, Texas, has researched thermal imaging of people's faces. This shows increased blood flow around the eyes, nose, and mouth when people are stressed, especially when lying. His research took thermal videos of thirty-nine political activists who were given the opportunity to steal a check for a small amount of money, left in an empty corridor, made payable to an opposing organization. Each subject was filmed picking up the check and then, during the video, asked if they had done it. Some of them were caught lying, and thermal imagery showed better than an 80 percent reliability. Other methods of analyzing our emotions exist, and continue to be developed.[11]

Many of these systems are able to recognize six classic emotions (surprise, happiness, disgust, anger, fear, and sadness) 90 percent of the time, and are often able to hit fifty-fifty guesses like "positive" and "negative" 75 percent of the time. So we can call this operational if it's right, on average, three-quarters of the time. When it gets combined with other information, like tone of voice, body temperature, or (most important) text, this additional data helps bolt the analysis into place, ensuring that the emotion being measured by the software is the emotion being experienced by the person. After all, sarcasm or forced responses can be confusing, even for a human.

All of these are just technologies that develop emotional profiles, which will be used to build emotional user interfaces. After all, most, if not all of this work, is based on measurement and the ability to abstract it into algorithmic descriptions.

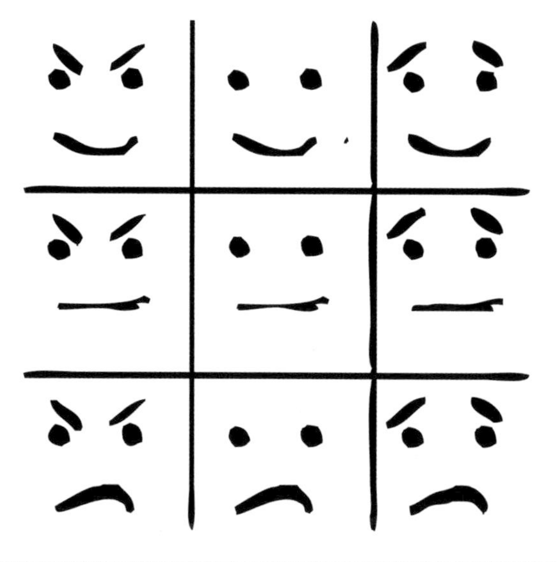

Mark Stephen Meadows

[10] *International Journal of Wavelets, Multiresolution and Information Processing*, by Karen Schmidt and Jeffrey Cohn.

[11] Keep an eye out for laser Doppler vibrometry, which measures tiny stress-related changes in respiration and heartbeat from afar.

After more than a decade of work, Rosalind Picard of the Massachusetts Institute of Technology published a book titled *Affective Computing*. She said that technologies would work better if a user's feelings were taken into account. A software tutoring program might change its pace or give suggestions if a student was looking frustrated, just as a human teacher might.

One of the results of this research was a device called an "Interactive Social-Emotional Toolkit" (iSET), which was designed to help children with autism, for example, or other personality traits that had been linked to processing emotional and social cues. The system works by monitoring the face of someone the child is talking with, and tracking the person's facial movements during the conversation. The face is then classified into one of six states: disagreeing, agreeing, concentrating, thinking, interested, or confused. Next to the face comes a pop-up screen that helps the child determine how the other person is feeling. This technology was then used to help the kids learn to identify emotions for themselves. It was an emotional dashboard to help special-needs children navigate the emotional landscape of the adult world.

Picard said that technologies would work better if a user's feelings were taken into account.

Picard's peers accepted her research by rolling their eyes and sucking their teeth. Her critics claimed that this wasn't "real research," calling it "soft" or "too fuzzy." They claimed it didn't address the hard sciences that robotics had to tackle. But luckily for Picard, opinion, like the industry itself, continued to change, and her work has since been accepted as the next step for robotics in the coming decade. Artificial emotion and social-emotional interactions are stepping stones robots are now crossing.

One example of robots implementing specific approaches is CB2 (Child Robot with Biomimetic Body). Since it was first invented at Osaka University in 2007, CB2 has been given emotional skills to interact with humans by watching their facial expressions. The research team, comprised of neurologists, psychologists, and other experts, has been teaching the android to respond to humans by evaluating facial expressions and then classifying them into basic categories, such as happiness and sadness.

The 130-centimeter (4-foot, 5-inch) tall, 33-kilogram (74-pound) robot has built-in eye cameras that record expressions and user data. It then uses this data to improve interactions with those same individuals later. The robot can also match expressions with physical sensations, which it detects via 197 pressure sensors under an exterior of soft silicone skin that, personally, I would have a hard time touching.

"Kismet" is another example of a robot capable of expressing artificial emotion. Dr. Cynthia Breazeal, a postdoctoral associate, also at MIT's Artificial Intelligence Laboratory, has believed since 2001 that emotions are key to social robot/human interaction. Programmed to learn and respond to its environment, Kismet was able to register and express a number of emotions, or likenesses, such as happiness, fear, and disgust. Sponsored by the Office of Naval Research, DARPA, and Nippon Telegraph and Telephone (NTT), Kismet was able to not only respond physically to people and objects in its immediate environment, but also to adapt to social cues such as rapid movement and gesture. As Breazeal put it, "... [Kismet] seems to really impact [visitors] on an emotional level, to the point where they tell me that when I turn Kismet off, it's really jarring. That's powerful. It means that I've really captured something in this robot that's special. That kind of reaction is also critical to the robot's design and purpose."[12]

Kismet was able to register and express a number of emotions, or likenesses, such as happiness, fear, and disgust.

Most recently, one of the single largest projects dedicated to artificial emotion in robotics was begun in Europe. Feelix Growing is a €2.3m research project that involves over two dozen roboticists, developmental psychologists, and neuroscientists from France, England, Denmark, Switzerland, and Greece. The aim, according to Dr. Lola Canamero, is to build robots that "learn from humans and respond in a socially and emotionally appropriate manner." The robots, relatively simple machines that look more like Roomba than HRP-C models, respond to tactile and emotional feedback. They more or less feed on words, behavior, and a person's helping it if it is physically stuck while trying to perform a task. What is most impressive is that the robots are able to detect facial expressions, gesture, tone of voice, and posture. From this, as well as other data, the robots detect emotional states such as anger, happiness, and loneliness, all of which are emotions that impact how the robot needs to behave.[13]

There are far, far too many projects to report on. Emotions and robots is an important field, and artificial emotion is a new frontier that is probably going to prove more fruitful than artificial intelligence.[14] Some of these many projects include the EU-funded JAST project, David McGoran's Heart Robot, The Kansei robot (created by a group led by Junichi Takeno, of Meiji University's School of Science and Technology), Maja Matarić's research at MIT, the EveR2-Muse, Space Robot, or NICT's body language robot, which can observe, recognize,

[12] For more, visit http://www.ai.mit.edu/projects/humanoid-robotics-group/kismet/kismet.html.
[13] For more, visit http://www.feelix-growing.org/.
[14] See the chapter on intelligence, below, for the reasons why.

Mark Stephen Meadows

The iCub is an ongoing project for a consortium of European researchers.

and react to human body language. There are many others, thousands of them.

But what do these developments mean for us in the next decade?

First, emotion-measuring systems will be used by law enforcement agencies in the coming decade. Many will be used in airports, taxicabs, train stations, and by police officers in towns that don't mind their presence. We will come to trust these systems more than our own perceptions, and as a result, our own perceptions will become less trustworthy. But these systems will continue to be adopted by robotics, and specifically android-based systems, as these will continue to evolve as a media platform for emotional engagement, and, as Dr. Hirukawa put it, entertainment.

It is through text that we reveal our innermost workings of love, desire, antipathy, and fear.

If androids are platforms that can be used for entertainment and health care, and my grandmother is in the hospital (being cared for and entertained by a robot that is covered under her health-care package), will I be more likely to visit her if she's getting emotional nourishment from some little robot seal that reads her face and purrs when it's supposed to? I hate to admit it, but I'd probably let the android (or the seal, or whatever-the-hell it is that's taking care of her), do the job for me. Or, when I can check in on my mobile device to see how my grandmother is doing, and probably talk with her through that robot, and I'm on my way home in the car, will I be more or less inclined to stop by to visit her if her pulse, words, and facial expressions indicate she's physiologically doing okay?

Probably less inclined. Robots that replace humans can be helpful in some contexts, especially in the dirty, dangerous, or deadly tasks, but when it comes to human-to-human interactions, the algorithms that measure our emotions, and the second-tier effects, such as leaving Grandma alone with her pet android, seem less-than-ideal uses of the technology. It's better than not having someone there, but I suspect that technology's ability to bridge gaps gives us a higher probability of being further apart.

After all, who wants to go to the nursing home to visit sick old Grandma when they could instead spend the afternoon lounging around with their sexy singing android?

How Robots Measure Your Emotions Via Your Words

IF THE SUPER BOWL OF FEBRUARY 1, 2009, GAVE US a contemporary and "reverse" version of a Voigt-Kampff test, it is nothing like what we saw during the following Super Bowl, of 2010. Nor is it anything like what has already been used to measure you. It all happens via text.

Text is an old, old technology. Enormously abstract concepts have wormed their ways through the millennia buried in only a few dozen letters. Text, like fire or fear, has a magic capacity to leap from one body to another. Imagine the musings of some Egyptian priest, once scrawled into a muddy wall, get transcribed to a parchment and sit for centuries in a tomb. The parchment is discovered, opened, transcribed again; it leaps into publication, appears in books, explodes to other countries, and then sits silent on shelves for decades. Later, it is again discovered, burned onto a hard disk, moved to an e-mail, and transmitted across the planet to appear on someone's brand-new telephone as they rocket across a Japanese countryside in a bullet train. The musings of that Egyptian priest travel through eons of history, suddenly to appear on the tiny screens we use on the metro today.

We can do a lot with only twenty-six letters of the alphabet. Text is the foundation for most math formulas, for all manner of classifying everything from plants to stars to blood cells, and is the foundation of software languages such as Java or C++. Text is the DNA of human knowledge, and each letter of the alphabet is a base, each punctuation mark a hydrogen bond.

Everyone uses text. As you read this sentence, millions of people at their office desks, on street corners, in cars, in the middle of the Pacific, on the icy bottom of the planet, in Rovaniemi or Rome, in Ciudad Juarez or Cape Town, are staring at little screens, worrying or smiling as they look at a text message they've just received. As text strings flicker across our planet's surface, we determine how to live our lives and what to think and feel. Text transfer, such as messaging, remains the primary value and main use of the Internet. It is the core input/output valve for human hearts around the globe. And it is through text that we reveal our innermost workings of love, desire, antipathy, and fear.

Those text strings are full of that magic material we call emotion, and like tiny train cars loaded with an invisible elixir, each letter carries emotions to the person on the other end of the line. Then, arriving at its terminus, the train is emptied, the cars are reshuffled, and it is sent back. You give and get emotions. You send text, you get some back. You send more text, you get more back. Sometimes it is objective data, and sometimes emotions get intense: surprise, anger, happiness, etc. Nothing guides our lives and determines what we do more than the text interactions we have. I'm trying, but it's damn hard for me to overstate the importance text has had in human evolution, culture, and technology.

We all use it, and it is the primary interface for all data today. Everything rides on it.

With text machines in hand, let's return to the Super Bowl advertisements, now 2010's Super Bowl advertisements, specifically February 7, 2010, during Super Bowl XLIV.

Google showed the following search strings in their Super Bowl ad:

study abroad paris france
cafes near the louvre
translate tu es très mignon
impress a french girl
chocolate shops paris
what are truffles
who is truffaut
long-distance relationship advice
jobs in paris
AA120
churches in paris
how to assemble a crib

With this list, Google managed to explain, in thirty seconds, how the story of someone's life can be assembled from their search queries. "Search on," concludes the advertisement, as if to say that Google can be used in every aspect of a person's life, from getting the right education, to marrying the right girl, to raising kids. That's true: You are what you Google.

Google learns about your life from the text you send it. The Gmail service, for example, gives you two gigabytes of free storage. So you write e-mails and leave stuff in there, and then, as soon as you're not looking (and sometimes when you are), Google scans those messages to figure out what you like and why (remember, Google is an advertising platform—it's just doing its job). Google also uses "cookies," which can link which pages you've looked at with your interests, tastes, and dislikes. It also uses Desktop Search and Documents to look through some of your deeper interests. It also uses voice, image, video, and any data that you send in or out of that system. It also works with other companies, like Apple,[15] to collect voice data, position data, videos, and photographs to register what users are seeing, and listen to the conversations that are spoken or heard. All of it goes into a big vat of data and each piece is individually tagged.

The commodities of the Internet today are online services in exchange for personal information.

By using various flavors of semantic processing,[16] Google (or Facebook, or thousands of other online services) quietly sifts through any material you've sent via their server and analyzes it. If you send a text message, or click on a link, or read a particular page for a particularly long time, the system notes this. It collects all the tiniest details, all the things you think are unimportant. It is as if there are a thousand tiny bees that descend upon your head, lick up your data-sweat, and pluck from you the small hairs and dead skin, taking those samples and returning to their nest where they can then determine what food you like, based on what you had yesterday, so that they can offer you something better tomorrow.

[15] Specifically, the iPhone or iPad.

[16] I'm speaking quite generally about latent semantic analysis, which uses natural language processing to compare various forms of data and to draw generalizations from them. It is also used to search for specific words, ideas, or recurring patterns, like "jihad" or "patriot" or "paranoid," for example.

The commodities of the Internet today are online services in exchange for personal information. Google gives you a mail account, and you give them a little insight into what you're thinking. It's a bit like letting your landlord read your mail in exchange for a discount on rent. Or maybe, more precisely, Google is more like the land barons of fifteenth-century Europe. They are land holders that charge peasants to live on that land. Peasants work there and Google charges some tax, but not too much, so that those people using their land feel it is appropriate to pay—just enough so that Google continues to gain. This is how Google harvests the intellectual property of the people tilling their virtual real estate.

Whatever the analogy, you contribute to Google's value. Google learns about you from the text you send it and from text it sends you.[17] When you run a couple of searches like "What is a robot" and then you run another search like "What is the definition of a robot," it refines the search results and notes what you are looking for. This data then goes into the big cache that Google uses as it develops search rankings. So Google learns not only from what you send it, but from how you respond to what it sends you. The search algorithm is a deeply interactive loop which can be customized for each user of the system. This loop, we hope, creates something good for everyone.

Ideally, as is the case with all advertising, this will create another loop of supply and demand that is in perfect balance. You giveth to Google and Google giveth unto you. Sellers find buyers and buyers find sellers. The notion is that search-and-advertising (the hypothetical opposites of one another) implements a program of balance and exchange, and at the end of the day, everyone goes home happy, healthy, and wealthy. Like all technology based on automated manufacturing, this, at least, is the dream. But the dreams of inventors plow paths of disappointment, so we should pay close attention to Google.

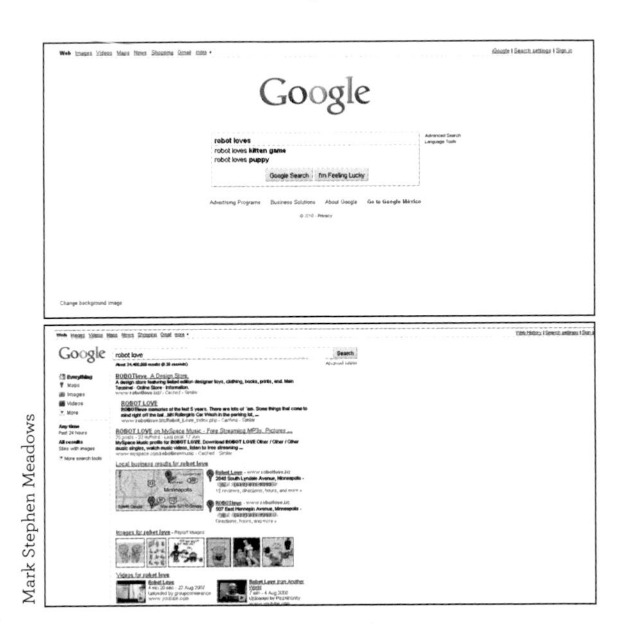

Mark Stephen Meadows

I bring up Google because it is a robot. And a damn strange one, at that. And as with any robot,

[17] It is a strange exercise to go to Google and type in simple strings, but don't hit return; wait for the tab completion to appear, and you can see what other people are feeling. Examples include typing, "E" and then waiting, or "St" and then waiting, or "How do I get" or "How do I learn" or "How can I be," or the time-wasting "How can I [insert one letter here]"—what amounts to public psychology spelunking.

it is being run by humans, and is therefore prone to human errors, whims, politics, and changes.

Obviously, it's one of those virtual robots, a knowledge worker, since it's not assembling cars or working physically. It's a robot that spans the various industries of service, industrial, and personal robotics. Of course, this depends on how we define the mythical word *robot*, but we can find a majority consensus even in the strictest definitions used by roboticists today. Search engines have an *input* and an *output*, they have *sensors*, they have *actuators*—they have some *degree of autonomy*. Any of the definitions of *robot* that I have played around with match what a search engine is and does.[18]

Google is a virtual robot, a polite one, with a fetish for semantics.

Google knows it is a robot and looks for instructions addressed to it. Just as if I were to arrive at a friend's house and find a note on the front door that starts with my name, Google does the same thing. It looks for a file called "robots.txt." This is a file that has traditionally been used to guide Google (and most other search engines) through a Web site. Google looks for this file and opens it up to see what permissions it is allowed. Google is a virtual robot, a polite one, with a fetish for semantics.

Many companies are working in this space of semantic processing and advertising. With Jupiter/Forrest Research forecasting the online advertising market continuing to grow by over 20 percent, annually, and Piper Jaffray predicting the industry could reach $55 billion over the next four years, this race will continue to accelerate. Software sold under the Sentry and FamilySafe brands is able to read private chats conducted through Yahoo, AOL, MSN, and other chat services and send back data on what kids are saying about movies, music, friends, and video games. The information is then sold to businesses seeking ways to tailor their marketing messages to kids. Adknowledge, another company that specializes in semantic-filtering technologies, sells online ads on a cost-per-click basis that's based on behavioral targeting and automated customization. The CIA uses semantic filters to locate terrorist hubs.

This is potent juju. What you want, what you think, and what you will do can be predicted with a 90 percent rate of accuracy, and that includes where you are now and where you will be in twenty-four hours.[19] In addition to your movements, Google can also predict your desires and emotions using these same

[18] International Organization for Standardization / ISO 8373 definition of robot is, "an automatically controlled, reprogrammable, multipurpose, manipulator programmable in three or more axes, which may be either fixed in place or mobile for use in industrial automation applications." Note, also, if you like picking nits, that an actuator may also be the read/write head on a disk drive. The word *robot* seems to offer little help, doesn't it?

[19] For an explanation of this, please read on, and see the February 2010 issue of *Science Magazine*, in which researchers looked at customer location data mined from cellular service providers. The authors of the study found that it may be possible to predict human movement patterns and location up to 93 percent of the time.

methods. Semantic filtering is not a terribly mysterious or hidden technology, and, in fact, it was widely used prior to becoming software. There's a huge body of open-source research on the psychology of semantics,[20] such as the Minnesota Multiphasic Personality Inventory (MMPI and MMPI-2), Myers-Briggs, Rorschach Thematic Apperception, and others, but they all basically stand on the work of Louis Gottschalk and Charles Osgood.

These methods of syntactical analysis can be used in computer programs to determine which words, for example, are most used, what they're related to, and in what context.[21] Here's an example, which is a slice from an e-mail I sent this morning (the numbers represent the relative frequency they have in the text):

1550 |

97 | i
96 | you
94 | Paris
90 | romantic
73 | we
72 | she
49 | the
41 | it
41 | but

Here's another slice, from about a tenth of the screenplay for the film, *Blade Runner:*

2065 |

98 | he
97 | Deckard
95 | Rachel
88 | i
65 | it
47 | the
45 | she
39 | you
25 | but

The same words were used with nearly the same frequency, but a couple renegades were in there which a computer program would immediately see as meaningful, and therefore likely to contain some emotional content. The difference provides the information, and the context provides the emotional measurement. This is just Psychology 101.

So if we can measure the word choices, patterns, and frequencies in a body of text, we can begin to build an analysis of it for meaning and emotion. Just as an analysis of cell-phone data shows where people were, and where they'll be next, emotional text analysis shows the same thing. We move across a daily landscape of emotions just as we move across a daily landscape of city streets. Semantic analysis (or, more accurately, probabilistic latent semantic analysis, and other technologies that look at

[20] For others researching the psychology of semantics, see the work of Erik Erikson, Carl Jung, Otto Rank, Gordon Allport, George Kelly, Abraham Maslow, Carl Rogers, Viktor Frankl, Rollo May, and Jean Piaget.

[21] Bayes's theorem, or Bayesian formula, being one of them.

our trends) shows where we'll be and what we are likely to think and feel. Just like predicting where people are from where they've been, we can also predict what people will think and feel based on what they've thought and felt. We humans aren't 100 percent predictable, but we *do* have a tendency to repeat our behaviors. Semantic filtering is part of a number of intense technologies that are intensely valuable. With tools like these, Google can analyze a user's data and discover the story of their life. What we search for indicates what we are thinking, what we are doing, and what we plan on doing.

In early December of 2007, Google's Marissa Mayer, then the vice president of Search Products and User Experience, showed how Google Trends can be used to predict presidential elections. She showed how Google Trends accurately predicted George W. Bush's win over John Kerry in 2004,[22] and also pointed out how Trends accurately predicted Nicolas Sarkozy's win in May of 2007, over Segolene Royal in the French elections.

The following year Trends also mapped the collective and accumulating interest in Barack Obama, who then went on to win the U.S. presidency in 2008. Other similar examples can be found by using systems such as Facebook, MySpace, or Twitter. Data from iPhone use is one of the most valuable sources of publicly created information, as it allows Apple (and Google) to collect data on what iPhone users are saying, hearing, seeing, and where they are in physical space. These systems generate enough data to predict movie sales, candy and magazine sales, and television show popularity, among millions of other data points.[23]

These trickles of personal data add up to rivers of valuable stuff. The data is so valuable, in fact, that it is becoming a major contention point between Apple and Google. Gene Munster, a Piper Jaffray analyst, says Apple will likely develop a search engine specifically tailored for the iPhone within the next five years. Since Google is the search provider for the iPhone, Google can see what iPhone users are searching for, which it then can use for its own mobile platform. The contract Apple struck with Google is effectually helping Google become Apple's main competitor, if not adversary.

Google channels the biggest river, collected from all the trickles of private data, into the mouth of the most insane, obsessive, giant, octopus-armed, spider-herding, pack-rat robot that the world has ever seen. Google discovers everything, watches everything, makes a copy of everything, and drags it all back to its nest, where it gobbles, digests, and ruminates. It sniffs around in our virtual homes and virtual mailboxes and virtual underwear, scans our blogs, reads our discussions, reviews the

[22] Note that not only were the overall victors correctly predicted, but the percentage by which the victor won was predicted as well: http://www.google.com/trends?q=bush%2C+kerry&ctab=0&geo=all&date=2004&sort=0.

[23] See the work of HPLabs, at http://www.hpl.hp.com/research/idl/, or their PDF at http://www.hpl.hp.com/research/scl/papers/socialmedia/socialmedia.pdf.

shopping lists we leave with various online services, reads our love letters, and determines what we like, what we don't, and when. It catalogs our emotions and motivations and keeps that data as a means of understanding who we are on a very intimate level—more intimate, even, than if our bodies were being measured. It makes a fingerprint and a catalog of us, storing the data out of our reach.

And why should we care? Why don't we like strangers looking through our e-mail or chat logs? Because the more information someone has about you, the more control they have over you.[24] Personal data is the fuel that powers tyrannies, police states, and dictatorships, because personal data is what allows personal control. The Medicis, the Habsburgs, and the Soviet KGB all knew that scanning letters was their means of maintaining government control. So does the American CIA.

Privacy is liberty. It allows you to think, do, and feel as you choose without risk to yourself or others, and this issue will become increasingly important as robots enter our homes and have a chance to observe us on a daily basis.

To return to the concept of hardware robots (and how virtual robots will integrate with them in the coming decade), here's a question: If Google offered you a robot to do your dishes, clean your clothes, and walk your dog, and it was as free as your Gmail account, would you take it?

If you already use Google's search engine, have a Gmail address, and use GoogleDocs, or use an iPhone, then you'd probably trust Google enough to give it a try.[25] So you'd click "Sign me up!" and ten days later a white box with the primary colors of Google shows up on your doorstep. You pull the box apart and pull out your new friend. GoogleBot, an android that comes up to your knee, just a little guy, with big feet and a cute head. You set him down and he begins to explore the house. Sure, for the first few days it was a little disturbing as it waddled around the house and read the books on your shelves and ran optical character recognition on all of your magazines, and seemed to be studying the contents of your refrigerator and dog food labels and clothing labels, and it very carefully watched your television, but eventually it settled down and stopped, and you think, *That was kind of cute*. It seemed to pay careful attention to your television and your music. Once you saw it touching your computer. Then, finally you got used to it. It did the laundry and fed the dog and Google got one more cluster of valuable data about your life.

This concept's now a prototype, and you might think that I'm about to point out that

[24] And to once again refer to Linda Stone's term, *understanding workers*, it is worth noting that as information becomes knowledge, and knowledge becomes understanding, your behavior and thoughts quickly become predictable.

[25] It seems conceivable that most iPhone users would be willing to pay for the product, since iPhone is serving a similar function—at least as declared in the Terms of Service.

Google is developing it, but that's not the case. It was some Swedes. Though iRobot swatted the ConnectR remote-controlled domestic robot in the start of 2009, ending the product line, the concept refused to die. A year later, in March of 2010, the concept was picked up by a group of Swedish hackers who launched a project similar to what I outline above, named GååeBot (pronounced "GoogleBot"). These Roomba hackers dissected the robot and inserted various items such as a camera, a Wi-Fi card, a small computer, and a good squirt of the AJAX programming language. After sewing it all back up, the thing still maintained operations as a normal Roomba.

But wait, there's more. It's not just a fly on the wall. The GååeBot can also read. It comes with optical character recognition (OCR) software so that you (or someone else) can photograph and translate text that happens to be lying around the house, drop it into a database, and run searches on that text. Source code and instructions are available for download.[26]

Google might consider giving these robots away in exchange for personal data (and users might consider the value of each). Google might, if they are serious about organizing the world's data, begin to move toward robots that link to products which are also RFID-tagged, connect with databases that track usage statistics, and can also read our emotions via gestures, facial expressions, and movement patterns. Google does not need to exchange only virtual services for personal data, as they do now; they might also decide to offer robotic services in exchange for this data.

Frankly, I thought this was a crazy idea. While working on this book I've asked three robotics professionals about it. Only one of them[27] replied, saying, "It will never happen. People would never want to be spied on in their own homes."

"But it's not spying if you invite it in," I explained.

"No, it won't happen. E-mail is one thing, but not with hardware."

We were sitting at a table in a conference room in his office, and as he said this his pocket made a beep. He pulled out his iPhone and began poking at the screen.

I was stunned.

"Do you know," I asked, "that most of the text, images, sounds, and any other data that goes into or comes out of that piece of hardware in your hand is being used by Google, or Apple, or one of several other companies? Take a look at the Google Maps section in the End User License Agreement.[28] Did you know that?"

He understood the question but still disagreed with my premise. We left it for time to tell.

[26] http://www.gaaglebot.com/.

[27] Bruno Maisonnier, CEO of Aldebaran Robotics, the company that builds Nao.

[28] http://store.apple.com/us/browse/home/iphone/terms_conditions which happens to also include http://www.google.com/privacypolicy.html and http://maps.google.com/help/terms_maps.html.

We're All Deckards

IN THE COMING DECADE, MOST OF US WON'T HAVE Paro seal-androids that are taking care of our grandparents in Scandinavian hospitals while a GoogleBot observes us getting undressed to mark our clothing brands. Most of us will, however, run searches on the Internet, and continue to exchange free services for personal information. Those services will become more granular, more personal, and closer to us. Systems and services that perform emotional analysis, prediction, and processing will assume intimate roles in our lives. Mobile devices will bristle with automated and autonomous sensors that will keep track of what we feel, do, and think. These systems will measure our thoughts and feelings in clever ways, register our likes and dislikes in remote databases, and build emotional models of who we are, how we think, and what we plan to do. These robotic systems, working invisibly and tirelessly, will build models of each of us.

We will call these robotic systems *user accounts.*

Each time some small bit of data about yourself goes out, each time you trade personal information for an online service, you become something different and you influence a system much larger than you. You become a part of a robot, perhaps one named Google or Yahoo! You contribute your emotions to the larger system, and that system becomes, in part, you, and you interface with a large robotic infrastructure that then tests someone else. It is not a Voigt-Kampff test. It is more than that. It is a Faustian test.

Deckard, we learn toward the end of the book and the movie, is a replicant. Ridley Scott, the film's director, has confirmed this in a few interviews, saying, "... Deckard is indeed a replicant. At the end there's a kind of confirmation that he is."[29]

In the final scene of the director's cut, Deckard leaves his apartment and a unicorn is left on the table. It is a sign that his memories of unicorns were imprints, and that those imprinted memories are available to Gaff, another Blade Runner who is now assigned to start hunting Deckard. So Deckard's emotions and motivations are known before even he, himself, knows them. Deckard doesn't even know he's a replicant. Deckard was able to measure the emotions of other replicants, but he never suspected that his own emotions were being measured. He never even knew he was a robot.

In *Blade Runner*, The Voigt-Kampff was a test that a human gave to a robot to measure its emotions. But this isn't how things are shaping up for the world today; instead, it's the robots that are measuring us.

And we humans are the robots.

---

[29] *Ridley Scott: Interviews*, edited by Laurence F. Knapp and Andrea F. Kulas (University Press of Mississippi, 2005).

Chapter Six: Star Wars

On Superstars, Masters, Slaves, and the Unholy Coupling of an iPhone, an Avatar, and a Cat—Interaction, Part 2

> The fact is that civilization requires slaves.
> Human slavery is wrong, insecure, and demoralizing.
> On mechanical slavery, on the slavery of the machine,
> the future of the world depends.
>
> —Oscar Wilde

I SHOULD PROBABLY SEE A COUNSELOR, AND HERE is at least one reason why.

I was at home one morning, at around ten, and I had called my bank's 800 number. My wife looked up at me as I repeated into the phone, in a monotone voice, "Four, four, nine . . . three, three, two . . ." I was gritting my teeth. I had been through almost an hour of this.

My wife sighed, looked back at her paper, and took a sip of her coffee.

I was none so peaceful. My morning had been lost in the muddy waters known as "customer service." *Customer service*, as you've probably discovered, is an informational moat that protects the castle known as "the company," and in that moat, "the company" throws all manner of awful robots that, like alligators, are there to prevent you from entering and talking to a real human that lives in the castle. It was dawning on me that the innocuous little chatbot named Alan—the one that I'd developed for Oracle in 2000—was here to take its karmic revenge.

Robot Rage

"I'M SORRY. I DID NOT UNDERSTAND," THE CHATbot said perkily. "Please repeat your account number . . ."

The bank's phone robot was telling me what I could and could not do, and how, and how not, to do it, and all with such a false tone of politeness (sarcasm?) that I had to try hard to visualize the human actor that had done the voice-over for the recording. I couldn't picture a real human, only a Barbie doll. It was uncanny. It was also grinding me down. So this "toll-free" number might not have cost me money, but this interaction was definitely taking its toll, and in the last twenty minutes I had grown tired. The phone robot, however, had not.

"I'm sorry. I did not understand," it said again. "Please repeat your account number." It pretended to be humble, and kept apologizing, but it was a completely domineering system. "I'm sorry. I did not understand. Please repeat your account number . . ."

I once again repeated my account number, this time, more slowly. I discovered I was sounding a lot like a robot, myself. The thing monotonously repeated my number, so slowly, and my options, again, and slowly. I was given more options. I was also only allowed to speak when spoken to. I could not ask it a question. I could not have a conversation, or even begin to explain that a transfer had to get split between two accounts. I was feeling weirdly oppressed, but in a subtle kind of way, like when someone holds a single index finger up to tell me to be quiet, yet smiles at the same time.

"Customer service representative!" I announced.

My determination made my annunciation all the more clear. Surely that would work.

"I'm sorry. I did not understand. Please press four or say 'Cancel transfer.' "

The robot was not sorry at all—it was just repeating what it had been told to say. It was not sorry because it was not capable of understanding. My head got hot.

"CUSTOMER SERVICE REPRESENTATIVE!!" I shouted into the phone.

My wife looked up again as I shouted, "HUMAN! GIVE ME A HUMAN!!" I put my hand on my forehead. "GIVE ME A FUCKING HUMAN . . . a customer service representative. A human. Gah . . ."

"I'm sorry. I did not understand. If you would like to make a transfer, pl—"

I banged the mouthpiece of the phone on the table, hoping the phone call actually *was* being "recorded for quality-control purposes," and then I lifted the receiver back to my mouth. I was about to scream something about spilling hot coffee on motherboards when I gathered my patience, put the phone back to my ear, and carefully declared, "Agent. Customer. Service. Representative. Human. Agent."

My wife continued reading, holding the cup a few inches off of the table.

"I'm sorry. I did not understand. If you would like to make a transfer—"

I hung up.

I walked over to my computer. Maybe I could chat with a customer service human via the Internet. I logged in and typed my way over to the correct page, clicked, and filled in my name. A little chat box appeared. It was just like any chat box. A string of text immediately followed, and it said, "Thank you for contacting customer service! My name is Dan. I am very pleased to have you as a customer. How can I help you?"

It was too fast, too well-typed, and too polite. It was the same robotic, impersonal, perfect kind of attitude that the empty bucketheaded robot on the phone had had. I said, "How many fingers do most humans have?"

"I'm sorry?" came the response.

"I'm testing to see if you are a robot." I typed. "Are you?"

"E chu ta!"(How Rude!)—E-3PO and C-3PO

PROTOCOL DROIDS[1] LIKE C-3PO ARE, AT LEAST IN the fictional universe of *Star Wars*, a line of androids specifically designed for social interaction.[2] With the intersection of far-flung cultures interacting on a regular basis throughout the galaxy, it's often necessary to have a communications specialist around. Protocol droids are there to help. They're designed to assist diplomats and politicians, to serve as administrative aides, and to accompany high-ranking officials on their delicate business.

Manufactured by Cybot Galactica on the factory world of Affa, the 3PO-series protocol units were the most advanced androids on the market. Their specialty was "human-cyborg relations." This, and the thirty other subtasks they were capable of performing, established them as the best-selling line of protocol robots across the galaxy for over a century. In response to the market success, Cybot Galactica began spinning off multiple other product lines that never quite nabbed the same market dominance (such as the TC- and 3PX-series). None of them were as popular, or as useful, as the C-3POs.

By contrast the E-3POs (which looked almost identical to the C-3POs, but were instead silver) came with a preinstalled TechSpan I module, which allowed them to interface with the Empire's many proprietary networks and various undocumented Imperial technologies. This exclusivity eventually gave the E-3PO line something of an inflated processor and caused them to be notoriously rude to the C-series 3POs.[3] The E-3PO was essentially, then, an upgraded C-3PO unit with a TechSpan card installed. What gave them some popularity also caused them to often be rude, even to their human owners. Eventually many of the E-series were later resold after the module had been un-installed, essentially downgrading them to standard C-3PO units and making them, again, modest and polite as a robot *should* be.

The C-3POs never had this problem with interaction. They were self-effacing, kind, loyal, adaptable, well-mannered, a bit squeaky when things got inconvenient, but as trustworthy and diligent as the best of their human owners. Though C-3POs could mimic any sound that was audible, they tended to default to an accent

1 It should be noted that the word *droid* is a Lucasfilm Ltd. trademark. Circumstantial evidence suggests it comes from *android*, but this is not technically accurate, since androids are shaped like humans, and R2-D2 is a droid. We'll leave Lucasfilm to its own vocabulary and avoid using this term unless addressing the fiction.

2 Ralph McQuarrie, the artist that initially imagined much of the *Star Wars* universe, relied on Fritz Lang's Metropolis for inspiration.

3 Much of this and the following comes from the great *Star Wars* opera, of course, but some confirmation comes from the honorable and ever-volatile Wookieepedia.org, where I refrain from so much as considering a contestation.

Lucasfilm / The Kobal Collection / Hampshere, Keith

Anakin got his start as a slave, and C-3PO as the slave of a slave.

reminiscent of BBC English. All this for about the price of a used landspeeder.[4]

Once upon a time a particular C-3PO model ended up in a scrap heap on the desert planet of Tatooine. Those parts were gathered up and patched together in a Mos Espa slave hovel, by Anakin Skywalker himself. The young genius managed to find enough of what he needed to build a rather solid robot, which was intended to help his mom, Shmi Skywalker, around the house. When Skywalker was finished, Threepio was fully functional and fluent in over six million languages (thanks to an onboard TranLang III Communicator module that was coupled with an AA-1 VerboBrain), and thus began the illustrious career of this most famous of Cybot Galactica's protocol products.

C-3PO got his start as a slave.

C-3PO got his start as a slave.

A communications expert, he was designed for interacting with humans. Built by a slave, and working as a slave, he helped out with the domestic work around the house. His first career was not only as a slave, but also as a family member (albeit, of the lowest order). He did his job, and evidently did it well, until his real mission as an ambassador and intergalactic translator finally began. The fussy android then graduated to become nobility, earned a big medal, and was finally remembered as one of the most important droids in the Galaxy. Quite a meteoric rise.

What I find odd about this story is how it is being played out in the real world today.

Say Hello to Honda's ASIMO

BACK HERE IN REALITY, WHICH IS ALMOST ALWAYS stranger than fiction, we have a very strange robot named ASIMO.

ASIMO is an international star. He has appeared on news programs on BBC, ABC, and NBC. He has famously conducted the Detroit Symphony Orchestra. He has been seen during the commercial previews in movie theaters, depicted walking silently through small Italian villas. He has competed onstage with Taiwanese children to see who could stand on one leg the longest. He has been to Edinburgh, Geneva, Moscow, New York, Lisbon, Taipei, Sydney, and Cannes.

ASIMO is one of Japan's UN ambassadors. He was the first "non-human" to ring the bell on the New York Stock Exchange. In October of 2008, ASIMO welcomed the Turkish minister of trade and industry to the Istanbul International Auto Show. In February of 2009, he

[4] I have conflicting information on canonical credit of the *Star Wars* worlds. I offer my apologies as board-game costs, RP game costs, movie costs, and Extended Universe costs each differ, so a credit crunch in the *Star Wars* Universe seems evident; $17,000 for a landspeeder, $10,000 for smuggling passengers into Alderaan, indicates to me that this price is off. What seemed even more certain to me was that a droid would not cost as much as, say, basic Trooper armor, but probably twice what an EV-9D9 Interrogation Droid would run. A DH-17 blaster is nowhere near as valuable, in utility or technology, as a protocol droid, so I hope my relative interpretations are acceptable to the culturally educated reader.

made a guest visit to the *Times*, in London, to show off his stunts to journalists. In November of 2009, ASIMO opened the Futuro Remoto Science Festival in Italy. In December of 2009, at Windsor Castle, he offered congratulations to prize winners. And in January of 2010, he made an appearance at the Sundance Film Festival to promote his new film, like any international celebrity.

More people have seen ASIMO on television, movie screens, and Internet videos than any other robot—maybe as much as all the rest, combined. He is one of the oldest humanoid robots, and certainly one of the best designed. Over the years, as is the case with any great celebrity, his look has been modified by his managers. His movements have been refined, his dialogue has been honed, his dance moves improved. His stage presence, personality, and outward appearance have all been gradually, professionally, and studiously polished by his managers to better please the palate of a mostly American public.

To say that ASIMO is a media star is an understatement. He not only makes appearances in the media, with plenty of bookings and appointments, but he also has a full marketing team managing his tours. ASIMO is paid for these appearances, and quite well (or, at least, his management team is paid for them). He is an *idoru*.[5] He represents an entire class of new media. This is why he's a media representative for an entire industry. ASIMO, like the president of a country, provides a public face, a media presence, and a polished political stance to an entire nation-state.

ASIMO is the ambassador of the robot kingdom.

But where to see ASIMO? After all, like some political dictator, he has body doubles everywhere. There are almost three dozen ASIMOs around the world, each one conducting orchestras, reaching up to take new toy cars, shaking the hand of some politician, waiting to be controlled by a BMI. Like some Hopi deity, such as the fox, you can never really see ASIMO, only his instantiation. There are many, but only one.

Where could ASIMO's home be? The question plagued me for many months, until, eventually, thanks to a friend and LA native named A. J. Peralta, I found my answer in a press release dated June of 2005. It was a small press release, nothing of great aplomb, a curious entry in which two virtual worlds seem to have strategically surrounded and silently sandwiched our own, physical world.

It reads:

> *CALIFORNIA, USA, June 1, 2005—Mickey Mouse today welcomed ASIMO, the*

[5] *Idoru* is an English transliteration of the word *idol*. It is a term that science fiction author William Gibson popularized in his 1997 novel with the same name, as an expression of a synthetic or fabricated media personality that may (but probably does not) exist as a real human.

world's most advanced humanoid robot, to a new home in the Honda ASIMO Theater, inside Disneyland Park. Guests visiting the popular Innoventions attraction at the park can now see ASIMO in an all-new live science show, Say "Hello" to Honda's ASIMO, opening June 2 as part of the theme park's 50th Anniversary celebration.[6]

Located in the middle of Los Angeles, the center of the galaxy of celebrity stardom, it was surely the best place to see ASIMO. It is where the highways and boulevards of Los Angeles somehow, by unmeasured gravities, pool, intersect, and disgorge the resident populations of Southern California into a kind of media cesspool.

Disneyland is a human delta, a kind of flood zone of flesh. Traveling in family-like tribes, these peoples are disgorged into a collected swamp, and we with them. The flood, when we step into Disneyland, flows around us and we merge with it and become part of the mess. Carriages filled with round people bobbed and buoyed in this inundation, the babies in carriages becoming undifferentiated from the obese adults in wheelchairs, all of them swirling and blending together into an hallucinogenic parade led by Mickey and his baton. Behind him came Minnie and Goofy, constantly dancing. These three rulers of Disneyland conducted the children and the elderly, like three pied pipers, toward the Honda Theater, in Tomorrowland, near the heart of the amusement park.

Commanded by robot voices in the trolley cars and park entrances, we were told to follow. And follow we did, A. J. and I, accompanied and enmeshed in this great tide of humanity. The huge Americans, slowly ambling through the park or rolling in their wheelchairs and baby carriages, jalapeño-cheese pretzels in hand, dressed in drab blues and reds and wearing ball caps and backpacks, surrounded me like a rising tide of polyunsaturated fat. To Tomorrowland we waddled.

After some time being lost, we walked up to the entrance of the little theater, inside Disney's Innoventions Center, where A. J. quietly said, "Look!" I don't know what I had expected, but I did not expect a little velvet rope to guide us into the theater. Two of ASIMO's biggest fans were sitting on the floor in front of the ASIMO theater, so we joined them and waited. We four were like fans at a rock concert, sitting in line, waiting to get the best possible seats.

Having just gotten off the Space Needle, I was feeling a little giddy, so I told myself that it was okay to sit down on the floor like this. As we sat there, the four of us each looked quietly at the other and I wondered what was motivating each of us to such limits. I looked at my watch. We still had twenty-two minutes to go.

---

[6] http://world.honda.com/news/2005/c050601b.html.

Mark Stephen Meadows

ASIMO comes in after having fetched the mail for Soccer Mom.

Soon a small woman came out, opened a door, and we filed into a small cinema with a large curtain covering what seemed to be a stage. On the overhead monitors Disneyland trivia was displayed, followed by small questions, quizzing us. My watch now said ten minutes till showtime. I looked at A. J. and he seemed less than pleased. But eventually (no one else was coming into the theater), a monotone and robotic voice over the loudspeaker announced we had five minutes left to wait. A short history of robotics was played, and then a short history of ASIMO, and then some two dozen people all came in as if on cue. The music started.

The show hadn't even begun, yet it was already so tightly coordinated, so rehearsed, so immaculately conceived, that it became unbelievable.

The curtains parted and an energetic and mildly attractive woman in a cardigan trotted out onstage and welcomed us to "her home." She extended her arm and rotated slowly to show the riches of her imaginary household. She seemed extremely excited.

She told us she was a soccer mom, and she had just sent "Awe-simo" out to get the mail and newspaper. Her neighbors will be impressed, she said. And when her family gets home, her husband, kids, and her parents, as well, will all love the new family robot. Her life, she told us, will be easier, more fun, more leisurely, and it will provide her with more time to spend with her kids. She looked out the stage door onto what she imagined to be her front yard, and she was as excited as a little puppy as she said, "Oh, oh, oh . . . HERE HE COMES!"

ASIMO walked into the room as she began to clap, rapidly, her elbows held close to her ribs. The audience clapped, too.

The uncanniness of the woman receded as ASIMO stepped onstage. I was soothed by this smooth little robot as he gracefully entered the room. He walked a bit strangely—it was a bit uncanny—but he was quite adorable. It was easy to see why he has achieved such celebrity. He was charismatic; cute enough to be harmless, smooth enough to be modern, stout enough to be durable, and his broad shoulders gave him the appearance of also being dependable. He had it all. Honda's great success with this robot must be attributed to the product designers of the system's physical and visual appearance.

As with any android, the function follows the form. As long as he's good-looking, he can get away with any mishap or mistake. ASIMO is like Britney Spears. And surely ASIMO's charisma is equal to, if not greater than, Britney's. Plus, he won't age, he won't shave his head at inopportune times, and he won't get into legal fights over babies. He's a perfect celebrity: He'll neither get tired nor make complaints about his dressing room.

ASIMO is like Britney Spears.

I blinked and tried to focus on the details of the presentation. What I was seeing was a talk-show setting, as if it were *Oprah* or *The Tonight Show*.

At 1.3 meters tall (about 4 feet), and weighing in at 54 kilograms (119 pounds), ASIMO walks at an average human walking speed (1.6 km/h), runs at an average human running speed (2.5 km/h), and is able to use all five fingers to generate a grasping force of about a half-kilo. The battery lasts for a bit more than an hour, depending on what the system's doing. Sensors in the foot record six different axes of distance information, and there is also a gyroscope and acceleration sensor (among others) built into the robot's torso. The head, arms, hands, and legs contain a total of thirty-two servomotors.

The robot's look is similar to an astronaut's, which gives it a fine futuristic appearance. Even

Mark Stephen Meadows

ASIMO saying hello.

if the icon is from the 1960s, it still conveys the message of "visitor from the future."[7] The size was partly a pragmatic choice to allow ASIMO to operate freely in the average human living space. This size allows the robot to operate light switches and doorknobs, and work at tables and workbenches. Its "eyes" (or at least the appearance of such) are located at the level of an adult's eyes when the adult is sitting in a chair. Honda tells us that a height of 120 centimeters makes human-robot communication easier.

The researchers at Honda believe that a robot height between 120 centimeters and that of an adult is ideal for operating in human living space. What Honda's not saying is that a robot should be smaller than a person. We already have, at least in the West, enough of a fear formula built in when we see a humanoid machine, so making it smaller than us is important. "Less intelligent" should be on that list, too.

Masato Hirose, executive chief engineer on the ASIMO project of Honda's R&D company, points out in a corporate promotions video[8] that particular attention was given to the face, and that there was a need to put eyes where they were not necessary. The robot needed eyes at the bottom of its stomach, for seeing stairs, but this was not what a human is used to. Hirose also points out that the robot does not have a mouth. It was important that humans be at ease with ASIMO. "[I]n order to make a robot useful in the home we had to consolidate all of the technologies we developed into a small size and design which people could feel comfortable with."

7 I haven't been able to confirm this, but I've heard multiple people say that the engineers were inspired by Astro Boy, which makes excellent sense to me.

8 *Living with Robots*, produced by Honda.

After all, if robots are a mirror, we want to like what we see when we look at them.

ASIMO is essentially a remote-controlled *karakuri ningyo* designed to deliver the illusion of autonomy. The robot's name is an acronym for "Advanced Step in Innovative Mobility," and this gives us a great clue into not only the design goals, but also the future of this robot. ASIMO's essence, as his acronym suggests, is a robot designed first for mobility and second for service.

ASIMO is essentially a remote-controlled *karakuri ningyo* designed to deliver the illusion of automomy.

First, it needs to be pointed out that ASIMO is a remote-controlled system, with very little decision-making capability. The PS3 (ASIMO's progenitor) and ASIMO are both controlled by a remote, similar to one you might use to operate a remote-controlled car, or a television.[9] As with an avatar in a virtual environment or an online world, ASIMO has few autonomous functions on board, and the functions he does have are used for mobility. The interface for driving ASIMO, in fact, is very similar to an avatar's. High-level decisions (such as direction of travel) are given to the human, while the system has enough autonomy to cover the details of navigation (such as where ASIMO has to place a foot).

This division of labor—between autonomy and decision-making—is very important. The human makes the decisions, and the robot follows. It is a classic division that most of the robotics industry is using now. What we see here is that locomotion, physical system maintenance, and movement are all controlled by the robot, while "higher" decision-making is left to the driver, the human controlling the system. So, for example, if I want ASIMO to walk down the stairs and then go to the middle of the room to do a little jig, there are hundreds of steps involved, but just three primary phases.

First, the driver can see, via a video connection, through the cameras mounted in ASIMO's head. So it is possible to tell that the path is clear and that there aren't any large objects in the way. When the joystick is pushed forward, ASIMO's cameras can determine, based on contrast, that there's a step in front of the system, and the cameras can also determine the distance to descend. The system takes over, slows down progress, balances the weight onto one foot, moves the other foot forward, and starts the descent. This is very hard to do, even with current-day technology. There are a huge number of motors that need to be coordinated so that the entire robot can balance on one small post. Meanwhile, the human driver sits and waits. Like a rider on a horse, the operator observes while those calculations are happening and ASIMO descends the stairs.

9 http://world.honda.com/HDTV/ASIMO/.

Second, once we get to the bottom of the stairs, the robot can be preprogrammed to turn left and walk to the middle of the room, or, alternatively, the driver can do it manually. When the robot arrives at the desired destination, a third step happens—the dance—which is entirely preprogrammed. As with the HRP series, there are a number of 3-D models that are all positioned and timed—an entirely virtual version of the robot, almost as if it were the robot's avatar—and this animation is then played back, on-screen. Once the author of the dance animation is happy with it, the dance is uploaded to ASIMO. This software then coordinates all of the pistons, cylinders, and rods to make an entirely new kind of dance.

With the labor divided like this, the human is responsible for the symbols and the robot is responsible for the acts.

In 2005, ASIMO's primary upgrade was that the remote-control device became portable. Honda also added several key functions such as walking on an angle, turning in place, or cornering. Additionally, a number of prerecorded movements that allow the robot to wave with one or both hands, bow, or make a grasping motion were added.

So ASIMO's "walking technology" is a large collection of both pre-scripted and on-the-fly processes that help ASIMO anticipate what is coming and lift its foot to adapt. A turn, for example, requires a different movement than an angle or cornering. This technology has probably been the key advance, other than the visual design of the robot, which Honda has focused on. ASIMO has some autonomy it can be preprogrammed for recorded moves—but mostly it is "driven," just as you might drive a remote-controlled toy car. ASIMO is first an electronic puppet and second a mechanical person.

Honda is a company that remains focused on transportation. As Stephen Keeney, the project leader for ASIMO North America put it, "We're not really a car company or a motorcycle company—we're a mobility company . . . Robots are all about extending and improving human mobility. Through a humanoid robot, people will be able to experience a kind of vicarious mobility." Is he talking about a kind of avatar? Other members of Honda's public relations department have used similar lines, such as Jeffrey Smith, a spokesman for American Honda Motor Company, who said, "Honda sees itself as a mobility company . . . If a person who's in bed can say 'ASIMO, answer the door,' it gives that person a kind of mobility."[10]

This makes ASIMO a kind of vehicle. Part of this mobility technology toolkit is based on ASIMO's foot sensors. If a map is input into the system and the robot is instructed to move from point A to point B, and along the way there are obstacles big enough to be avoided, the foot

[10] http://www.financialexpress.com/news/little-visitor-from-the-future-can-walk-the-talk/42006/0.

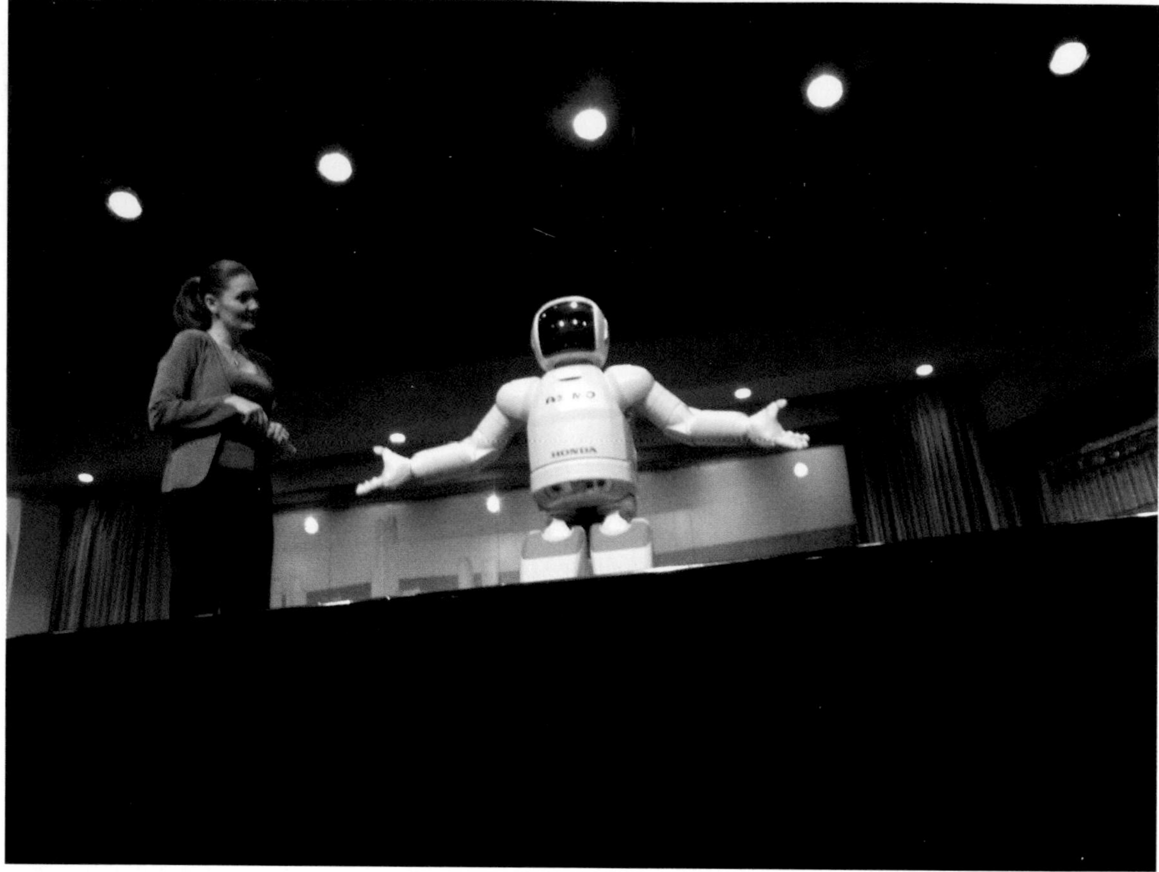

Mark Stephen Meadows

ASIMO! Superstar slave!

sensors and visual sensors can help it navigate around these things. In the wrist there are sensors that allow ASIMO to move if it is pulled or pushed, and visual sensors also allow it to move in conjunction with the people around it.

This strange vehicle also has object-recognition capabilities. Gesture recognition (such as being able to tell the difference between someone waving hello to you and someone flipping you the bird) is based on object recognition. As objects are "seen," angles are measured so that vectors can be resolved (such as where the hand is pointing versus where the finger is pointing), and these are then simplified into smaller forms. These can be related to spoken words so that ASIMO is able to learn what a fork is, or what a cube is, and then be able to refer to it later.

Despite this ability to learn what objects are, ASIMO is still confronted with one of the great hurdles that humans leap so easily, which is hand-eye coordination. Right now ASIMO is designed to move its hands toward an object to indicate interest, but grasping an object on a table, even something like a small cube, remains a major task, albeit one that will be overcome in the coming decade, certainly.

Of course ASIMO is connected to the Internet so that it can do things like answer "What's the weather forecast for tomorrow," transcribe voice-to-text messages, answer questions by doing site-specific lookups,[11] send MP3 files from one unit to another, and perform other things we normal Internet users do. And of course this connectivity makes downloading new maps and animations quite simple. But consciousness, intelligence, and the ability to predict and make decisions are all still attached to the human hands that run the remote-control device and preprogram the dance steps that ASIMO is performing on stage.

Even with such great advances, ASIMO remains a kind of *karakuri ningyo*—an advanced electronic marionette that is dedicated to service and human mobility.

"We're sooo happy to have ASIMO in our lives!"

As she faced the audience, the Soccer Mom pressed her hands together almost as if she was about to pray. She bent forward a little and exclaimed, "We're sooo happy to have ASIMO in our lives! He can do sooo many things for us!"

A. J. and I looked at each other and we both raised our eyebrows. We had just heard a strange haiku of the most curious sort emerge from this woman's mouth. With it, the ASIMO marketing team had coupled a person and a product.

To deconstruct a bit: When someone uses the phrase, "We're happy to have X in our life," it is, in my experience, typically used when referring to a member of the family. Amping up the importance with the word "so" has become something in modern parlance that is used to indicate how emotionally important that phrase is, such as "Thank you so much" (as opposed to, say, "Thank you very much"). It is intended to be personal and emotional.

Soccer Mom was *so* happy to have ASIMO as a new family member—one that was going to do things for her family.

This coupled an intense emotional closeness with what amounted to a slave-like appliance.

[11] As far as I have been able to gather, the ontological intelligence of the system is limited. It is not, for example, doing much more than reading the text that it accesses, so this means that no interpretation of the answer is being given, only a response to a query via something like Wikipedia or Google, without digging much deeper. I add this footnote because I was not able to confirm this information.

The robot walked toward her, shuffling along the polished wood floor of the stage.

ASIMO has a strange walk. It is clear that there's a large amount of balancing going on, an intense load of processing. He looked uncomfortable as he bent forward a little, keeping his knees bent and sticking his can out, as if he'd just crapped his carton.

Soccer Mom put her hands together again, this time clapping, and said, "Please help me welcome the newest member of our family, ASIMO!!" We all applauded again. Then Soccer Mom put her hands on her knees and looked at ASIMO with a sympathetic posture.

ASIMO was a member of the family, though not equal to the parents, grandparents, or children.

"ASIMO, was there any mail?"

"No," came the droning voice, childlike but still canned, "but your neighbor gave me this."

ASIMO handed her a piece of paper.

She squealed, "Thank you, ASIMO!"

"I'm an amazingly capable humanoid robot. How can I help you?"

Then, Soccer Mom called her husband on the videophone. He appeared in a wheelchair. She asked him if she could keep the robot, and he said yes. This seemed to be intentionally lifted right from *The Jetsons*. Soccer Mom hung up the phone and turned back to ASIMO, saying, "With your ability to interpret voice commands and navigate, it is easy to see how valuable you'll be, helping me around the house."

With this exchange, ASIMO has been clearly labeled as a household servant. ASIMO didn't get a chance to respond because the phone rang, and it was the kids. They screamed hello to the robot, more or less forgot about interacting with their mom, and made it abundantly clear that all of their friends would be coming over; the family would soon be achieving newfound popularity in the neighborhood.

The phone was hung up and Soccer Mom continued with the demonstration of ASIMO as he walked forward and backward, avoided objects, balanced on one foot, balanced on the other. She put a soccer ball in front of the robot and it kicked a goal (which made Soccer Mom particularly happy). Yes, ASIMO was a child, would be popular with the other children, and part of this was because he was a unique soccer player. But, as if in an act of self-control, Soccer Mom again pointed out that, ". . . his current role is to be a helper in the home."

She made it clear that ASIMO was a member of the family, although his social level was not equal to the parents, grandparents, or even the children. He was there to reinforce the family structure, the messaging seemed to say, but it would be by serving. He would be somewhere in between a child, a butler, and a slave.

"Excuse me," ASIMO's prepubescent voice blurted out from the speakers in the theater.

Mark Stephen Meadows

ASIMO pulls a few dance moves.

Soccer Mom looked at the audience with an expression balanced between surprise and offense, as if the robot—like any slave—did not quite have the right to speak unless spoken to. Or was it that he had exceeded his technical boundaries? She blinked rapidly at us, then slowly turned to face the robot, her hands again on her knees.

"Yes, ASIMO?"

"Would you like me to order the pizza for you?" ASIMO's genderless, youthful robotic voice asked.

She stood up and turned again toward the audience, tilted her head to the side, smiled, and put her palms in the air. They ended by celebrating with a dance, which they performed in unison. The poor woman. She was surely programmed, like ASIMO himself, by the marketing team.

To wash some of the weird out of our heads, A. J. and I went and got Cokes and jalapeño-cheese pretzels.

Outside, in the sun, things seemed a little clearer, although the message was no brighter. This notion of ASIMO being "an important part of the family" and a "helper in the home" is a new kind of marketing. It is no longer just that Honda is selling a product, nor even a lifestyle, but it's now selling indentured servitude. They seemed to be selling child-slaves. The used robot salesman from *The Jetsons* had no such sneakiness. Jane was not out to buy love from her robot; Rosey was a servant, plain and simple. And while ASIMO was also a servant, this idea of him being a family member seemed almost as important. It was as if we had just run into a woman on the street, selling marionettes, and she thought of them as her offspring, but as warriors and servants, too.

"If we don't want a robot uprising, maybe we shouldn't be making a robot slave class."

Was Soccer Mom playing out some kind of submission/domination fantasy?

The interaction had been intensely complex.

A. J. looked at me and, with his mouth full of pretzel, muttered, "If we don't want a robot uprising, maybe we shouldn't be making a robot slave class."

The Medium of Masters and Slaves

ASIMO AND C-3PO HAVE STRANGE CONTRADICTIONS in common, and it has to do with how they interact with humans. It is a slave/master relationship, and not one that seems healthy or simple.

First, both are communications experts, cut to fit into the highest levels of society. ASIMO is a superstar marketer. C-3PO is a human relations protocol expert. They are practiced at the arts of seduction, both of them knowing when to step forward and usually giving the appearance of acting in a supporting role. Whether it is marketing or translating, both are grand wizard communicators, capable of communication magic. They're both PR pros, and protocol droids (in proper *Star Wars* parlance).

Second, in that they have achieved elite status in the human world, they are masters over humans. They have spent time in the highest halls of justice and politics. Threepio serves his ambassadorship in the Royal House of Alderaan, with Princess Padmé Naberrie of Naboo, and at the Imperial Council. ASIMO, as one of Japan's UN ambassadors, has done his share of public service, too, in Turkey, England, and

Italy, to name a few. Both androids are highfliers, socially, orchestrating the interactions between different cultures. Or, as Honda puts it on their Web site, "Indeed, the positive responses to ASIMO as . . . ambassador are universal and cross cultures and languages."[12]

Simultaneously, in contradiction, these two robots are slaves. Both are domestic servants of the lowest order. C-3PO was built in a slave slum in Mos Espa. When Soccer Mom tells ASIMO to bring in the paper, he's treated with all the dignity of a dog. Both robots are treated as slaves and expected to maintain a humble and subservient position. They both function in the home as mother's little helper, keeping an eye on the kids while they do it. They're a bit less than butlers since they're not paid. Which means they're slaves.

So ASIMO and C-3PO are each simultaneously master and slave. They are each designed to interact with humans on the lowest level, as slaves, and to facilitate interactions at the highest, as politicians.

This odd contradiction is mostly due to the marketing choices for ASIMO. Surely the marketing group is waiting to see how the robot is best received, but it also seems clear to me that we don't yet really know what our relationship with robots—and with androids, in particular—will become. Will they be our servants, or we theirs?

I didn't know whether to pity Soccer Mom or the audience, but the schizophrenia ran deep.

Interactions with robots in the coming years are bound to get weird; in fact, they already are.

Interactions with sexbots are certainly a tour through the thorny gardens of both slavery and mastery. If sexual liberty in the '60s was the first sexual revolution, and drugs in the '90s was the second (with things such as erectile dysfunction drugs, birth control pills, and others), then sexbots of the coming decades will most likely be the third. Plus, they might be fun. Consider sex with dolls such as Andy, of First Androids.[13] She will never get tired or bloody; she will never get lice, yeast infections, herpes, or warts. Sexbots will never be sleepy, flaccid, or uninterested. They will always be turned on, always ready for a roll in the boiler room with you or your friends, or both.[14]

The most laudable sexbot project, and the one that is closest to the vision of real robotics, is Roxxxy and Rocky, products of the company named TrueCompanion.[15] In an interview with ABC of Australia,[16] Douglas Hines, who was an artificial intelligence engineer at Bell Labs

12 This phrase can be found on multiple Honda sites in Japan, the UK, and the U.S.

13 Be sure to visit the photo galleries of http://www.first-androids.de/ or http://www.first-androids.org/.

14 This is a fascinating topic that I just didn't have room for here. David Levy goes into detail in his clearly structured and carefully defended book, *Love and Sex with Robots*.

15 http://www.truecompanion.com/.

16 "Meet Roxxxy, the Robotic Girlfriend," January 10, 2010, AFP.

before starting TrueCompanion, says, "She can't vacuum, she can't cook, but she can do almost anything else if you know what I mean . . . She's a companion. She has a personality. She hears you. She listens to you. She speaks. She feels your touch. She goes to sleep. We are trying to replicate a personality of a person."

Odd interactions introduce second-order oddities. There's the story of the man that abandoned his sex doll in Japan;[17] the RealDoll models with blue skin and elf ears;[18] the face massage robot;[19] or the sex club barker.[20] I've heard rumors, seen photos, or talked with owners of thalidomide babybots, amputee grandmabots, and one Osaka resident I met who owned a robot that owned a robot that owned a doll. These robots, and the love their owners have for them, might seem strange, but we should remember that there is no such thing as normal in the world of robotics.

Via e-mail, I asked Dr. David Levy, author of *Love and Sex with Robots*,[21] what he thought of our relationships with robots, and if he saw a master or a slave relationship developing. He replied:

> *In the coming thirty years I see robots remaining as slaves, but as time goes beyond that thirty-year limit, they will become our equals in many ways, and eventually our superiors, and, in some ways, our masters. An interesting analogy is the way that the parent-child relationship develops; initially, the child is the slave (if you like), but over time the child learns enough that, in many cases, eventually it overtakes its parents intellectually and thereby becomes, in some sense, their master.*

As we design robots and figure out how we'll interact with them, the real question becomes how we'll interact with ourselves.

We can see a fork in the road ahead: It's the question of master or slave. ASIMO and C-3PO each have personalities that point to this fork in the road. Both of these robots present the need to make a decision: Will our interaction with robots be one of slavery or mastery? As we look into the face of our robotic futures, the eyes give no clue. The eyes of ASIMO are almost hidden and offer no hints of emotion. They do not wink, blink, water, dilate, or twinkle. The eyes of our robots, be they our overlords or our slaves, are unreadable. Perfect for being either a master or a slave.

17 On August 21, 2008, a sixty-year-old unemployed resident of Izu, Japan, wrapped his 1.7-meter-tall, 50-kilogram silicone girlfriend in a sleeping bag, drove to a remote area in the woods, and left it there. The doll was found some weeks later and the man was fined for littering. For more, see www.pinktentacle.com.

18 See www.realdoll.com/ and view their catalog for more.

19 See the oral rehabilitation robot, named "WAO-1" and invented by Atsuo Takanishi of Waseda University and Akitoshi Katsumata of Asahi University.

20 In the Minami area of Osaka, a humanoid robot dressed in a sailor suit would solicit visitors to enter, a practice I also saw one evening in Shinjuku.

21 *Love and Sex with Robots: The Evolution of Human-Robot Relationships*, New York: HarperCollins, 2007.

"I, for one, welcome our new robotic overlords,"[22] goes the joke. Will we even have robotic overlords? Or, conversely, as A. J. said, when we were at Disneyland: If we don't want a slave uprising, maybe we shouldn't be making a slave class.

Real-world robots will be neither masters nor slaves. Robots will become a new interface for creating masters and slaves. Robots will simply be the new medium used by masters to make slaves. Just as the Internet, television, radio, vehicles, bombs, and guns have been used to establish master-slave relationships, robots, too, will be used for the same power dynamic. This power dynamic will be quite subtle, and will be based on how available robotic technology is, how well we understand it, and what it is used for.

Much of this power dynamic comes from how technology is distributed and shared, whether it is open, available technology, or whether it is held in the hands of a few. Machines should be open to their owners to change, modify, and control. All power should reside in the user. If not, it can become a device to tyrannize, control, and even enslave.

This is just another argument for an open robot operating system, some kind of system that is as transparent as it is available, so that when we have robots in our home they are not monitoring us any more than we are monitoring them.

We will not, in ten years, nor forty (despite some of the cheers of the Singularity crowd), be bowing down to some humanlike Strong AI-in-the-sky, nor will we have little silver robot-gnomes praying to us while they busily polish our shoes and clean up our dirty dishes. We will be neither slaves nor masters to our robots.

Robots will be the new medium used by masters to make slaves.

But some of us will be enslaved, and some of us will become masters because of robots.

Robots are not our new masters. They are our new medium. With it comes all the responsibility of an increasingly powerful interaction, ability to communicate, ability to convince, and ability to control.

This will largely be done by replacing a person with a robot.

My problems with the phone customer-service robot had to do with the fact that I was being tyrannized by a person I could not communicate with. I found myself in an uncomfortable interaction because I was being forced to jump through design hoops that I didn't want to jump through, and in a manner I didn't want to follow. I was being told to jump through hoops like a little circus dog so that my master, the bank, didn't have to spend their time (and hence their profits), on having their employees

[22] Ostensibly this idiom is from *The Simpsons* (season 5, episode #15, "Deep Space Homer").

serve me. A person had been replaced by a robot, and that changed the power dynamic.

Some person, somewhere, decided not to give me access to the bank employees. I had been made unequal, and put on an assembly line. The phonebot had forced an interaction that caused me to be the one waiting in line, not the bank employees. It was a design decision, and probably a cost-motivated one. The phone robot was designed to establish a power interaction that saved the bank money, and some guy in an office somewhere made those decisions consciously. It's not some great AI that's rumbling forth from the bowels of the Interwebs. No, it is some bossy-boss somewhere who's kicking around a geek at a desk and forcing him to make a system that can control you to save them money. It's the bossy-boss and his geek that will be the masters.

Fortunately, there are alternative interactions.

Nao, the Pet-Friend from Inner Space

THE SUN WAS OUT. IT WAS A WARM DAY, AND SOME dry leaves crunched underfoot. I was walking through Paris's 14th arrondissement, on the swanky south side. All seemed warm and summery. But there was a cloudy problem ahead. It had to do with the re-forming of friendships and families. If a robot like ASIMO is supposed to be a child-slave member of the family, what does this mean for the carbon-based members of the clan? This cloudy problem on the otherwise sunny horizon was a social problem. It had to do with our friendships, and how these change as we interact with robots. Surely the Internet has changed our friendships; surely mobile devices and other portable computers change our relationships with our friends and families; surely avatars and profiles impact our lives and interactions. If all of this is true, then robots will, too. But how?

Nao is as sophisticated as he is small.

As I walked down the street, I was concerned about what would happen to the family if we invited a child-slave into the home. What would the kids do with that? What does a person feel after spending some years with a robot? How does it change how we address one another?

As I continued walking down Raymond Losserand, near the Plaisance metro, I noticed a small family of four coming toward me. The two girls in the family were each carrying a doll. The dolls, as all dolls are, were synthetic friends.

I wondered if soon they would be carrying a Nao.

Nao is as sophisticated as he is small. This robot has many of the same sensory capacities that humans have. In terms of eyesight, the system is equipped with two CMOS 640x480 cameras that capture images at thirty frames per second. One is in his forehead, scanning

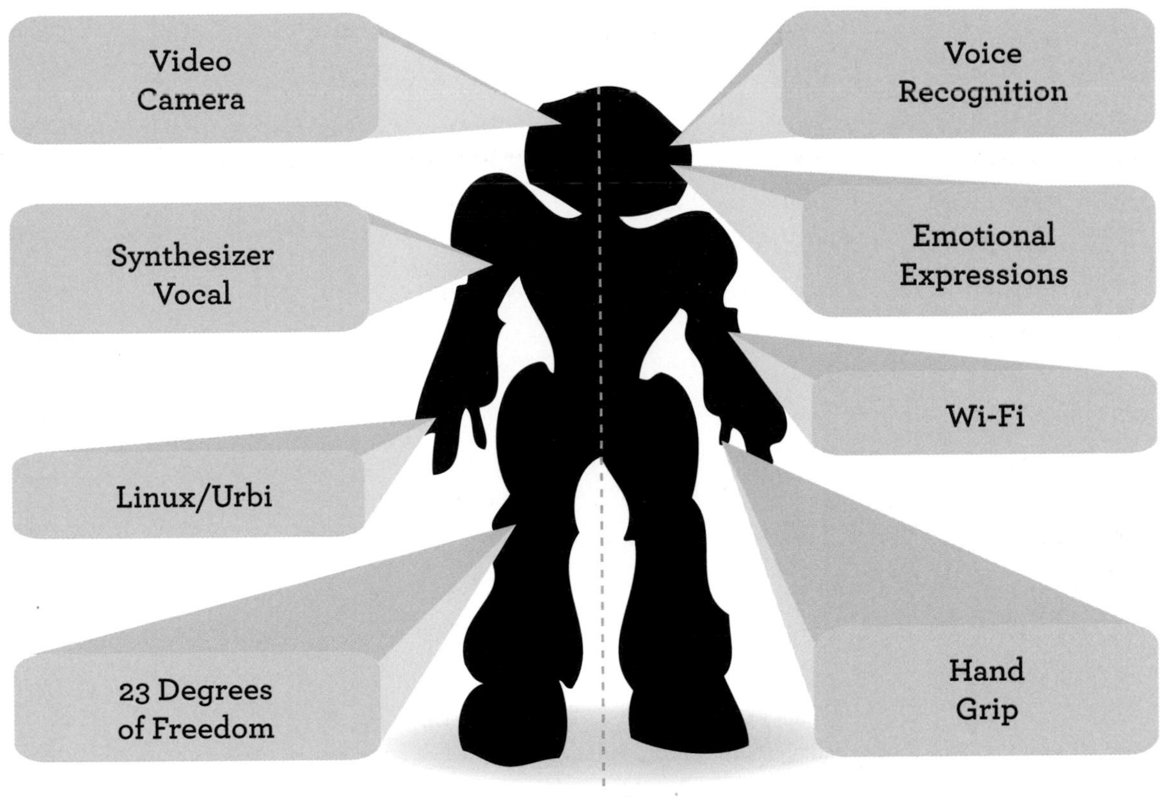

Overview of Nao's functionality.

the horizon above, while the other is attached to the bottom of his head, scanning the floor and walking space ahead so that collisions are avoided, and so that he doesn't fall down stairs, etc. These video recordings are stored and can be recovered for later review, but they are also used to develop a sense of the world around him. Object recognition, facial recognition, and the ability to perform optical character recognition all mean that Nao is able to see what's around him, understand it, and recognize it next time the same object is seen.

As well as objects, the robot can also recognize people by their faces and voices, and he can read text that is presented to his cameras. You can say "Go in the other room," just like you might to a dog, and he can do just that. Or, if your little brother gets a hold of it, Nao might throw food at you, instead.

For sound, Nao has four microphones, two speakers (with stereo capacity, of course), and onboard sonar capacities to help with navigation. He can read from a Web site in English or French via text-to-speech capability; sitting in

Mark Stephen Meadows

Aldebaran engineers at work.

your car with you, for example, as you drive to work, replacing the role of the traditional lector, he might read your e-mails to you, or he might just pipe in, directly, the MP3 of a friend's voice.

The robot has a clear sense of touch. With servos that operate in over 25 degrees of freedom, Nao is able to adapt to sliding or changing surfaces, correcting his balance as you might on a surfboard (with both an accelerometer and a gyrometer built into the torso, this skill comes as little surprise). Ultrasound transducers in the torso also help to keep the robot from hitting walls, and bumpers in the toes help Nao to avoid (or to target) kicking objects. Meanwhile, there are force-sensitive sensors built not into the fingers but into the motors, which means that Nao can pick up something fragile and not break it.

Consider ASIMO: During the demonstration in Disneyland, Soccer Mom put a soccer ball in front of the robot, and on cue, ASIMO lifted its foot and kicked the ball. Nao doesn't need such assistance. In fact, Nao, during the Robot World Cup, was seen not only running and kicking the ball, but also diving to keep it from going in the net.

So Nao has a wide array of locomotion capabilities. He also has a tactile sensor on the head, a *capacitive* sensor, more specifically, which is divided into three sections. This is something like the mouse we find on today's computers. It serves as an interface with the robot. Pressing once will turn the robot off, for example, but the tactile interface is also sensitive enough that complex animation sequences can be programmed into the robot via this interface alone. Normally the robot responds to being touched via animation, voice, and illuminated LEDs.

I went to Paris to visit the offices of Aldebaran Robotics, makers of Nao.

It is not a simple system. In terms of networking capacity, the robot has infrared, Wi-Fi,[23] Bluetooth, or he can read from Web sites, including Really Simple Syndication (RSS) feeds. With more moving parts than the average automobile,[24] the robot is controlled via a powerful onboard computer,[25] and the lithium polymer batteries are able to last more than ninety minutes per operation, after which he will go plug himself in and recharge, without being asked. Nao can recognize when other robots are in the room, and will immediately begin transferring information or coordinating movements, if told to, upon recognition.

These functionalities mean that Nao can play chess, read e-mail, wake you up, play soccer with other Nao models, read the news, collect things, distribute things, find things, greet you when you walk in the door, and then hide the things he just found when you weren't looking.

I went to Paris to visit the offices of Aldebaran Robotics, to see Nao and their workshop, and to talk with the robot's designers about Nao's basic design principles, how they differ from ASIMO's, how Nao's interaction with humans is different, and ultimately, to learn what cultural impact this technology might have on us in the coming decades.

The offices sit inside a former electricity storage facility, property of Électricité de France. Now a "business hotel," the building is owned by the city of Paris. Its facade is covered in ancient arches and tiny solar panels, where shiny steel has been set into the old stone of the hundreds-year-old structure.

I was sitting at a long wooden desk with Bastien Parent, Aldebaran's public relations manager. Bastien is a foxy fellow, with black hair, a

[23] 802.11 A, B, or G are possible.

[24] This statistic was provided by Aldebaran, who informed me that Nao uses over 1,300 distinct moving parts.

[25] X86 AMD Geocode 500 MHz CPU, 256MB SDRAM, 2GB flash memory.

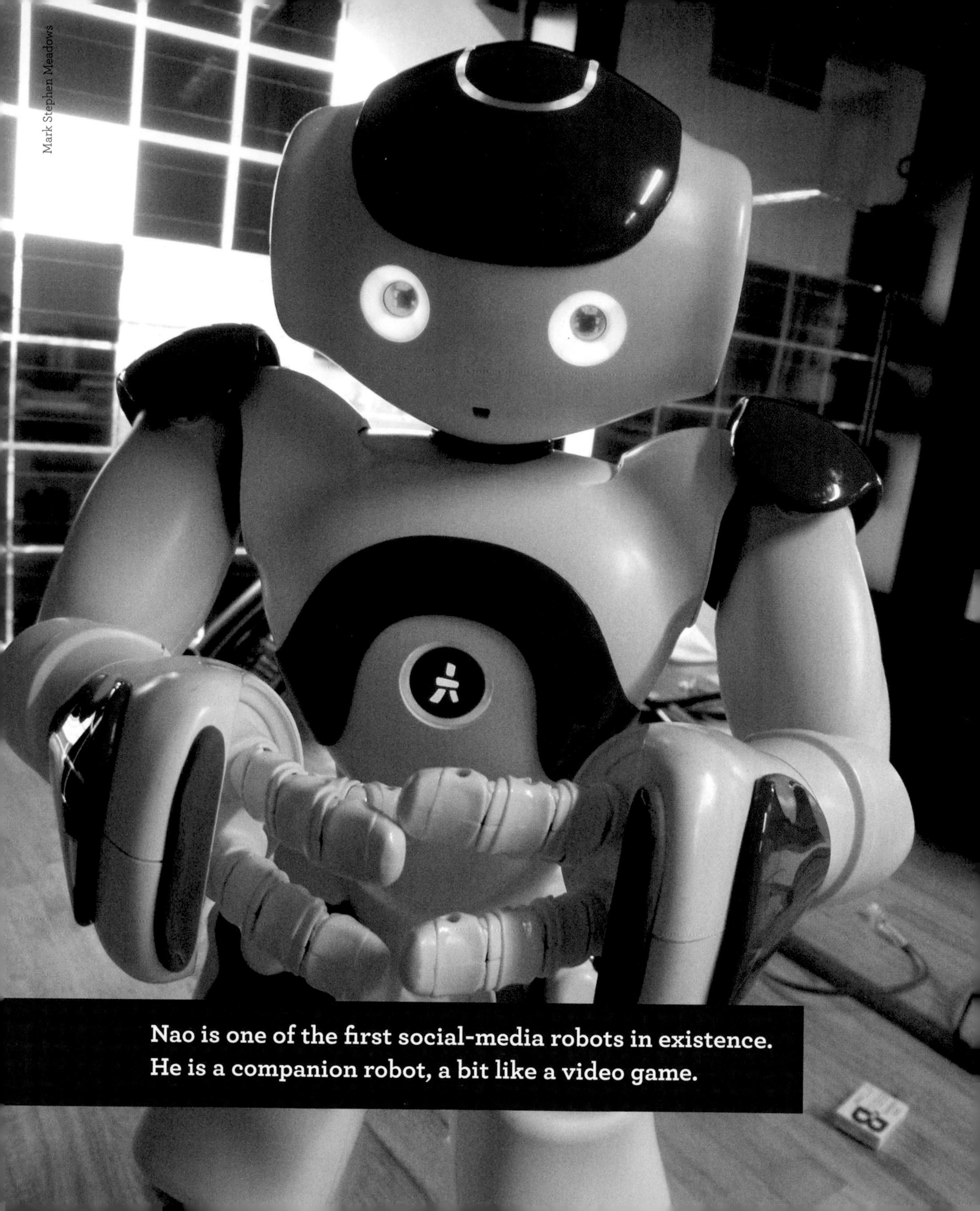

Mark Stephen Meadows

Nao is one of the first social-media robots in existence. He is a companion robot, a bit like a video game.

narrow face, and clever eyes. As the fifth employee of Aldebaran, he has seen the company shift and grow to twenty times that size. His blood and sweat is in the company.

He told me how Aldebaran got its start as the dream of company CEO, Bruno Maisonnier. Back in 2000, Maisonnier was working as a technology consultant, specifically for banks, and not too happy about it. He had been tinkering with robots his entire life, and even in the 1980s, he'd begun to wonder when the robotics industry would pull off a repeat of what the computer industry did. By 2005, Maisonnier had managed to convince enough skeptics that there was a future in the robotics industry; he gained the support of almost a quarter-million euro from angel investors—just enough to build a prototype Nao. He did some market studies, and worked alone for nearly six months until he was able to bring in two people to help. He acquired another half-million euro in 2006. With just a few grains of sand in the hand, by comparison to most American investments, Maisonnier had managed, by 2008, to put together a team of forty people who were well engaged in making what would later become Nao. By 2010, the company was nearly a hundred strong.

The shelves were neatly stacked, each containing different body parts . . .

The shop occupied the entire third floor of the building. Bastien slowly walked me toward one end of the building to start our tour. We walked into what looked, at first, like a warehouse whose walls were stacked high with metal shelves and boxes. In the middle of the long room were workbenches where people hunched over their tables, soldering little bodies, laid out like patients on open operating tables.

"This is our repairs and assembly room," Bastien said. "All the robots are put together here, and if one comes back, it gets fixed right here. It's artisanal."

What he meant by *artisanal* is that even though robots are electronic, Aldebaran still takes pride in the work, as if it were a handcrafted art, not a machine-made product from an assembly line. They see these robots as a doll maker might see their dolls, hand-stitching with proper tools, making them as high-quality as possible. They take pride in their craftsmanship and ensure that each model is solid and well built.

The shelves were neatly stacked, each containing different body parts: a hundred little feet, all lined up, toes out, in rows, like tiny shoes. Below that shelf were a hundred little arms, each about as big as a candle, each one bent at the same angle, each one lined up against the other like chevrons, and all of them, skinned as they were, showing gears and circuit boards and wires. And still below these were a hundred little heads. Two of them were oddly shaped and seemed to have hats.

"We were asked to equip these two with lasers on the head," Bastien commented with a sideways grin. Custom work. Each one designed to the highest standards the company could manage.

A couple of stock boys carefully moved boxes; a woman brushed her hair back out of her face and adjusted her glasses before carefully soldering together two boards. Another young man walked through the shop, carrying in his hands a defunct Nao, arms and legs limp, obviously a new patient for the little hospital. There must have been several thousand robots in the room, all told. I've never seen a robot manufacturing plant before, and it was smaller, friendlier, and happier than I'd imagined. It felt like a visit to some Italian marionette shop from the 1800s.

It felt like a visit to some Italian marionette shop from the 1800s.

"Once we have them finished, they get tested in here," Bastian said, indicating a large fish-tank-like room where little arms and legs were being pumped and spun, attached to testing motors and wires, like intravenous tubes, measuring voltage and amperage. The robots' eyes were dark, and I couldn't help but see them as unconscious.

As we continued through the offices, at each of the desks, people had a Nao that accompanied them in their work. One engineer with dreadlocks stared intently at his screen while his Nao stared intently at him. Without looking at his Nao, the engineer held up a red ball. Nao took it from his hand. The engineer continued, not bothering to look up. His Nao seemed like an avatar for him, one who had clearly moved beyond problems of hand-eye coordination. He was now engaged in mind-robot coordination. His robot responded to whatever the engineer typed into his computer, and it had to do with grabbing that red ball. Was it robotic hand-eye coordination that the robot was practicing via the human's mind-robot coordination? It was mysterious good fun, and the engineer was damn serious about it.

Further back in the shop there was a woman on a machine with a 3-D model of a head on her screen. "That's for Romeo," Bastien told me. "That will be a service robot we'll be releasing in a year or two."

"He looks larger. And ugly." I said.

Bastien laughed a bit as he replied, "Yeah, well, Nao is too small to push open doors, so if people want a robot that can actually perform larger tasks around the house, this is the model for them."

Many boogers of conceptual criticism have been flicked at Nao over the years. Most people are accustomed to robots that do physical things. People have asked, "What is Nao *good* for? If it can't even open the door, then what kind of robot is that?" Vacuum cleaners, for example, are the kinds of robots we are getting used to. Service robots are becoming a kind of industry standard. After all, the manipulation of the physical world is an obvious need we all have. None of us wants to do dull, repetitive tasks like vacuuming. There are robots for that, and Romeo will be one of them.

I suspect this is a response to a public reaction that has quickly overlooked what Nao is really best at, and that's social interaction. Nao is more of a physical, social-media platform than a vacuum cleaner. Nao is the unholy coupling of an iPhone and an avatar.

Consider what we do with cell phones, or most mobile devices today. I might send video or sound files. I might send both. I might, for example, use my phone to look up the weather, or get directions. Sometimes if I'm on hold, I'll put the call on speakerphone and do something else while waiting for a human voice to chirp on the other end. I've used my phone to keep track of my schedule and my contacts, and to record random thoughts that I'll jot down and store for later. I've used my phone to record audio, video, photos, store MP3 files, make to-do lists, and then to organize them. I've used my cell phone as a light in a dark hallway and as a doorstop.

Same with Nao, who can do every item listed above. Now, let's couple that with avatars. Consider what we do with avatars, or most online virtual selves that are humanoid. For myself, I have spent much time with them as adult dolls. I have spent hundreds of hours building poses, working on animations, selling those poses and animations, inventing new workarounds for people who don't have the know-how or patience to debug their systems, making clothes, skins, and accessories, and exploring the psychology of what it means to have a mirror of myself that does what I tell it to do. An avatar is a kind of self-portrait. It is a social version of ourselves, and something that we use to practice and test before setting out into the real world with real ideas.

Same with Nao. He can perform every task in the above list. What this means is that Nao is one of the first social-media robots in existence. He is a companion robot, a bit like a video game. But it's not like a video game from the '80s, which keeps you alone in your basement as you get fatter and zittier. It is more like a video game of the last decade, in which you interact with others, via a shared narrative. Nao is a bridge to other people, in addition to being an interface with other Naos that are nearby. It can be used not only as a networked communication medium—so I can talk with people remotely, and use it like a phone—but also as something that we can play with collaboratively, in much the same way as people play with avatars in World of Warcraft, or the world of Barbie. Nao is a companion that bridges to other companions.

It is also like a pet. It is small, slightly dependent, trainable, attentive, and displays emotions in response to touch or words. Fortunately, it doesn't need a litterbox.

In another coup, Nao is designed to be autonomous. It can charge its own battery, upload new behaviors, and has an embedded decision system so that it can live its own life and go about its own business. You can still control the robot with PC, direct interface, or cell phone (among other devices), allowing it to dance, collaborate, or fight with other robots, but the robot is designed to be autonomous first, and remote-controlled second. This is a big difference from other systems, such as ASIMO or the HRP models. Nao is less like a servant and more like a very independent and intelligent pet.

It's like the child of an unholy threesome between an avatar, an iPhone, and a cat.

"But if I have kids and they spend hours every day with this thing, does that make them geeks?" I rhetorically asked Bastien.

"Well, what's the case with other technologies? If you give a kid a Facebook account, does he become less social or more? We did some user studies and filmed families when they got their Nao, during the out-of-box experience. What we found was that every video was the same. It was like Christmas. The whole family was there. Everyone wanted to see the new robot. Everyone gathered around the box, and when the thing was opened and Nao started to move, everyone lit up with the robot! It was great to see, really."

Mark Stephen Meadows

Natanel Dukan and his Nao, more a pet than servant.

"And when the hype died down and the youngest members of the family started playing with the software?" I asked.

"The parents went out to dinner to celebrate. The social interaction and closeness continued, on new levels. Nao brings people closer. It gives the family something new to do together. It's like a new pet."

"Okay, but after a few years? Then what?"

Bastien put his hands on his hips and leaned his head to the side, his mouth slightly open. He blinked once. He looked upset with me for not understanding.

"Nao can do anything." He looked serious.

"*Anything?*" I asked. Horrible images of antigravity orgies, lava lamps, and hookah pipes appeared in my mind's eye.

"Yes. *Anything*. The system is open-source, so we're waiting to see what programmers come up with. We're relying on our community to lead us in the right direction with these applications. But there is no limit to what we can do here."

Shaking the hallucinations from my head, it seemed reasonable to imagine seeing-eye robots (retinal implants that interface with the robot for both remote and local interface), hearing-ear robots (cochlear interfaces), and robots that are able to help folks that are too old, too weak, or too injured to do things like pick up a fork that has dropped, or collect a bedpan.

Still skeptical, I asked Bastien for a demonstration.

We walked over to one of the desks tucked back in the corner, where Bastien introduced me to Natanel Dukan. On Natanel's desk was a red model Nao, motors in idle. Natanel flipped the robot on and we went through some various exercises that included dancing, singing, and other things much like what I'd seen HRP-4C do at the AIST labs. Additionally, there were some gripping and balancing exercises. Overall, although the system was noisy and the mechanics were less than clean, all the thousands of moving parts were working great, and the robot stopped itself before walking off the table, or over a piece of paper, or a dozen other gauntlets we threw down for it. Eventually, as we concentrated on talking and ignored Nao, the robot seemed to get bored, and he gently sat down, rotating his head to look at Natanel. Although Nao was small, he was the most advanced, most human, and most practical robot I'd seen.

As we talked, Natanel kept a caring eye on his robot, and at that point I couldn't help but ask him the same basic question I had asked Bastien about social interaction. I decided to take it to a slightly more personal level.

"What's your relationship with your robot?"

Natanel smiled and looked at Bastien, and Bastien smiled back. There was a shared experience here—the profound and subtle experience of identical twins or brothers-in-arms.

"He's like a pet," Natanel said. "I like having him around. Okay, it's not like I can't live without him, but if he's not around, I kind of notice."

"Like your computer?" I wondered aloud.

"No, no," Natanel said. "I really need my computer. My work and other things depend on having my computer near me. Nao's a pet."

Chapter Seven: Battlestar Galactica

On Resurrections, Autoportraits, Top-Downs, Bottom-Ups, and Why AI Is an Obsolete Art—Intelligence, Part 2

Their idols are silver and gold,
the work of men's hands.
They have mouths, but they speak not;
eyes have they, but they see not;
they have ears, but they hear not;
noses have they, but they smell not;
they have hands, but they handle not;
feet have they, but they walk not;
neither speak they through their throat.
They that make them are like unto them;
so is every one that trusteth in them.
—*Psalm 115*

IN THE COMING DECADE I DOUBT WE'LL SEE HUmanlike AI appearing anytime soon, but we might see some dead people get resurrected. *Battlestar Galactica*, and its prequel, *Caprica*, do a fine job of outlining how this could happen.

The 2004 television series seems to, at first, serve us the same sci-fi robot recipe we're used to eating. The recipe contains the following ingredients: humans build robots, robots build more robots (named Cylons[1]), robots try to kill humans, mostly succeed, struggle ensues, and soon robots and humans are indistinguishable. Interbreeding follows. The line between robots and humans disappears. And finally, *Battlestar Galactica* adds a biblical cherry to the top of the cake. It introduces through most of the series the notion of resurrection.

How to Conduct a Resurrection

I LOVE *Battlestar Galactica*. I LOVE THE WILD camera angles, the nutsy bravado, and the blonde actress who plays Number Six. I also love *Galactica* because of its commentary on today's world. Like *The Matrix* did in the 1990s, like all good science fiction, *Galactica* presents a snapshot image of the social issues that were happening when it aired. It commented on many ideas, including genetic engineering, politics, religion, and immortality through software.

[1] *Cylon* stands for "cybernetic life-form node."

Battlestar Galactica (Sci Fi Channel) Season 4, 2007-2008. Pictured: (l-r) Mary McDonnell, Tricia Helfer, Michael Hogan, Jamie Bamber, James Callis, Tricia Helfer, Katee Sackhoff, Michael Trucco, Aaron Douglas, Grace Park, Tahmoh Penikett, Edward James Olmos. Please do not overlook the reference to daVinci's painting, *The Last Supper.*

In *Caprica,* the prequel to *Galactica,* which takes place some seventy years before *Galactica,* we get to see how the Cylons first boot up. A successful engineer and all-around computer genius named Daniel Graystone builds a series of military robots. They're your typical clunky, dumpy, bolt-headed mechanical man-soldiers. Unknown to Dr. Graystone, his daughter, Zoe, who was fooling around in virtual worlds, managed to upload her personality to her avatar. As a result of Zoe aggregating her user data from a number of online sources, the avatar was acting, thinking, and behaving just like her—everything from what articles she wrote to what birth control pills she was taking. This avatar was an aggregation of her online activity, called Zoe-R.

Zoe (the physical Zoe, that is) dies in a blast of religious zealotry, an acolyte of a new religion that sounds quite Christian, or at least monotheistic. Zoe-R lives on. After Zoe is dead, her computer-genius father downloads her avatar personality and puts it on a kind of glowing hockey puck. He carries this around for a while, occasionally talking with Zoe-R, and, eventually, trying to hang on to the personality of his daughter, he installs it into one of his battle robots. The thing looks like the bulldog-faced Mack-truck robot it is, but inside it is trapped poor, little teenage Zoe-R (usually pictured in her chiffon baby-doll dress). Zoe-R is now the operating system of her father's robot. When Daniel boots up the robot, the Cylon's first word, spoken through thick digital distortion, is "Daddy?"

This is the oddest immaculate conception I've ever heard of.[2] It is also our first resurrection and the beginning of the Cylon species.[3]

Eventually, Cylon begets Cylon in a sexless conveyor belt of automated manufacturing, and as they polish their technique, the robots start to make themselves progressively more like humans. They manage to make themselves seem more like very sexy humans, in fact, and they come in twelve delicious models. There is an indefinite number of each model, like car models, and each model has its own personality and memories. The individuals have memories which can be shared with the group. So if one Cylon dies, it just transfers its memories to another member of its model.

This means that there's a personality database that Cylons share. It means that there can be many instances of one personality, and that instance can be copied, shared, and built on by any one of the Cylons. It's a kind of open-source artificial intelligence virtual personality, detached from hardware. The mind is separate from the body.

It becomes very hard to tell what the Cylons are up to when Number Six (the big blonde chick) manages to resurrect herself in the consciousness of one of the humans, an AI engineer named Balthazar. Balthazar is our iconographic Judas Iscariot, the betrayer of the human race, and so it is curious to see Number Six choosing him to be the only human that has access to this mind-in-the-sky. At any rate, *Galactica* goes on, during its years of airing, to show one resurrection after another, each in various forms, but always with the mind cleanly sliced from the body.

This mind/body split has been a big source of contention among AI developers over the years. Many people argue that the body and mind work together to create our understanding of the world.[4] Others argue that the mind is entirely separable from the body. The

---

[2] What I mean here is that Zoe equals God, Daniel equals Mary, and Zoe-R equals Christ.

[3] To be redundant, what's recounted here is not, specifically, *Battlestar Galactica*, but the prequel that takes places some six decades before *Battlestar*, called *Caprica*.

[4] Of course, the argument is far more complicated, and involves not only intelligence but also memory, emotions, and all of the other nonphysical characteristics that make us each individuals, but I'm swapping *mind* for *intelligence* here, and hope to use it as shorthand to avoid getting too deep into details around distinctions that will only interest specialists.

arguments flare high on both sides of the debate, but the consensus is, as *Galactica* states, that intelligence and personality and mind and emotion and all things nonphysical are separate, or perhaps separable, from the body. The meme is that intelligence, like soul, doesn't really need the body.

This meme is probably about 3,000 years old, and probably began when Plato started talking about the *eidos* (ΕΙΔΩΣ) as separate from the object. That way of thinking got picked up by the Judeo-Christian guys, who claimed that we had a soul and a body. Those guys passed it along to us moderns, and these days, even though we have software and hardware, we still think like Plato and the Judeo-Christian guys. So Plato handed it to Jesus who handed it to Bill Gates. Software's the new soul.

Plato handed it to Jesus who handed it to Bill Gates. Software's the new soul.

It makes sense if you consider your telephone answering machine. You pick up your telephone, type in a few keystrokes to tell the system to record, and then: "Hi. Sorry I missed your call. I'll get right back to you as soon as I can." Enter in a confirmation to save, then hang up. Something of you is then uploaded into that machine. Just enough of your personality is stored in a digital format so that when someone calls and interacts with that system, they get a little software equivalent of you. It's your tone of voice and your word construction, and when someone calls, they recognize these things and they speak to the telephone answering machine as if it's you (even though they know it isn't), and they ask it questions as if it will respond (even though they know it won't).

Your telephone answering machine is a very primitive virtual version of your personality. I'm not saying it *is* you, but I am saying it's a primitive representative of you, a little ambassador of a digital sort, stationed in the phone to behave like you the best it can to get your job done. And it's separable from your body.

Building virtual personalities has been the primary focus of my work for the last decade. I began some of this work at Xerox PARC, and later, at Stanford Research Institute, where the role of these agents could be used in experimental documents, interactive narratives, and video games. Later came Oracle, and eventually I started two companies dedicated to this rather peculiar kind of software.[5]

I'm going to outline some of the directions of my past work in hopes that the line between science fiction and engineering fact will become

[5] My personal goal has been to build a talking portrait, a kind of interactive *Mona Lisa*. Since I work with paints as well as with text and computers, the three of them always seemed to intersect for me, especially around portraits that were interactive and had something to say. It's always made sense to me that portraits should speak with words and gestures, not just with image.

more clear, and to point out what I think is the most important next step robotics research needs to take. I hope these details are helpful if only to point to some important online libraries and methods that can be used as we develop interfaces with robots.

One method of building virtual personalities is a bit like what's outlined in *Galactica*. It amounts to software for creating server-based conversational agents whose personalities are derived from an ontological knowledge base and customized rules created by users or derived from text samples. (Please see the appendix for more on this.)

My favorite example of our past work involved Arnold Schwarzenegger.[6] We collected several available interviews that Arnold had conducted in the last five years and fed them into a natural language and semantic recognition system we'd developed. Arnold's most commonly used phrases were flagged and sorted, and his most commonly used ideas were ranked in a similar fashion. Just as all of us do, there were some ideas he tended to use more than others, and some expressions that he favored more than most.

We asked the system, "What do you think of gay marriage?"

It answered, "Gay marriage should be between a man and a woman, and if you ask me again I'll make you do 500 push-ups."

Photo from NASA

Clustering is a common occurrence that follows mathematically discovered principles.

As if with an invisible butterfly net, we had captured part of Arnold's personality. It was him, but it wasn't. It had the components of his personality, but he had never typed nor said that particular string of one hundred and three characters. Much like a telephone answering machine, core elements of his personality were recorded and played back. It was identifiable as him, but it was not him.

It was an Arnoldbot.

The problem was, there were some questions we asked the system which just confused it. Part of the reason for this was because we didn't have enough data (we only used a dozen or so interviews). The solution to the problem was to use larger bodies of text. The frequency of some words is measurable, and those more-measurable words can be used to build a

[6] This was work we did at HeadCase, a company I founded in 2007, which developed much of this infrastructure. I worked with Paco Nathan and Florian Leibert, who did most of the engineering.

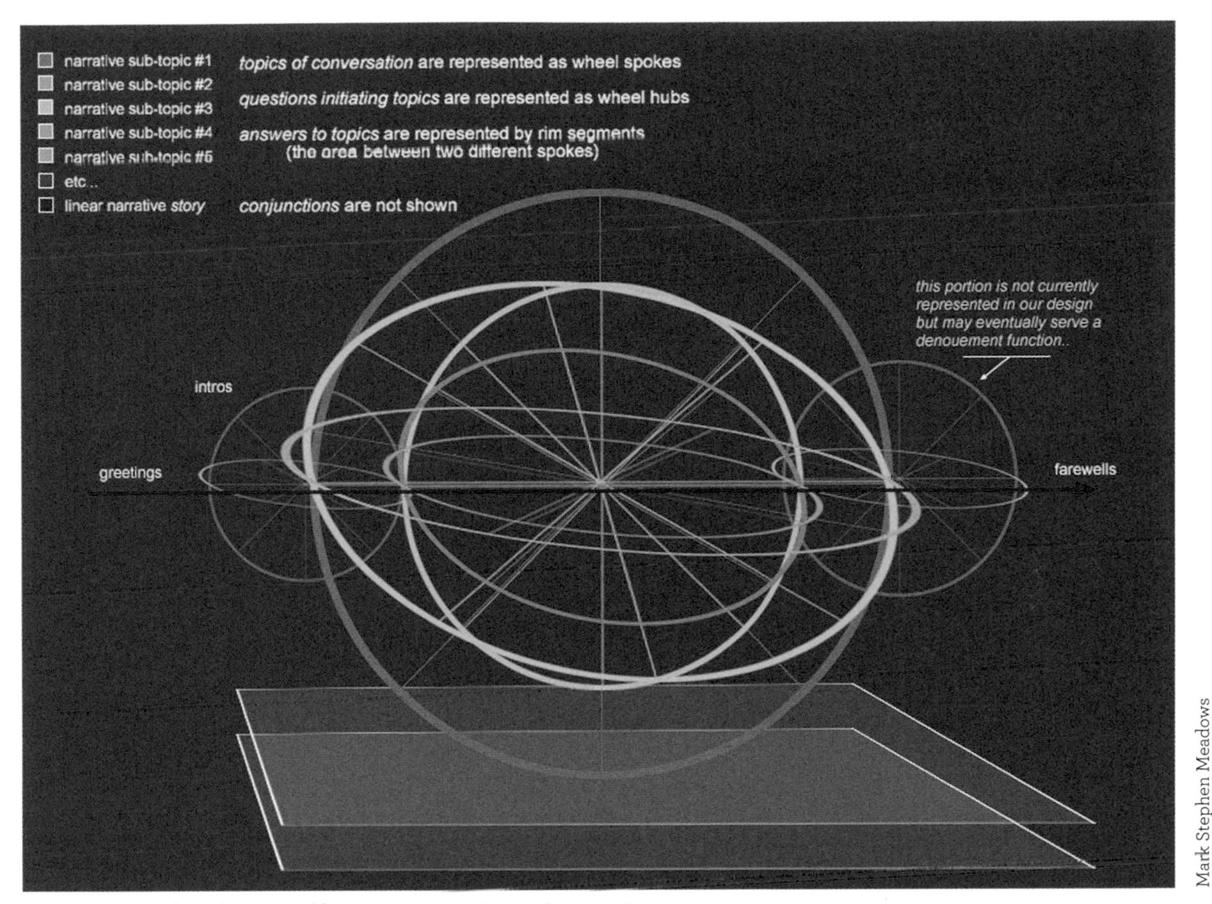

Mark Stephen Meadows

A narrative topology for natural language processing and generation.

profile, or a kind of fingerprint that is that author's unique text-based identity.

This has been suspected for about a century now. Linguist George Kingsley Zipf was one of the first to notice the possibilities, and he created a theorem called Zipf's Law (of course). This law states that some words are used more frequently than others, and that this builds a particular graph. The most frequent words are used twice as frequently as the next most frequent word. For example, the word *the* is used twice as frequently as *of*, which is used twice as frequently as *and*. And of the languages functioning in the world today, all follow this same pattern. This pattern makes a kind of spiral, a Fibonacci sequence, and common concepts and words cluster at the center; these can be mapped, like a city, with a kind of population density at its core. The vocabulary you use can be graphically mapped on this chart and compared to another person's, based on the words you each use.

In December of 2009, a group of Swedish researchers from the Department of Physics at Umeå University[7] took several bodies of text from famous authors and discovered that authors used words with particular frequencies and patterns. Those word frequencies and patterns recurred, per author, and the researchers argued that this created what they called "an author's fingerprint." Of course, what these researchers really meant was not that the physical bodies of these authors had identifiable fingerprints, but that their body of text did. There was a kind of personality to the body of text that was based on the frequencies with which some concepts were used (the index of a book is an example of this kind of thing).

This method can be used with blogs far more easily than with books, since they're already digital and copyable. Blogs contain a body of very specific word use, context, and conceptual structures that can be used to develop virtual personalities. The approach is basically the same as the Arnoldbot approach we used at HeadCase.

This method can be used on your e-mail, too.[8] After all, whatever you send in e-mail is a kind of intimate literature about your life, tiny snippets of personal journals that record what you are thinking, and almost always in the first person. Google and others (Yahoo! or Facebook or Twitter, among many others), while not composing a virtual personality of you, do use that information to build a personality profile, which is just enough for them to accomplish the job they need to do, which is usually to help you.

With e-mail, text messages, Facebook, Twitter, and blogs, we're all authors today.

If an author's personality can be extracted from their writing, their personality lives, in some sense, in their writing. It also exists online in user accounts, personality preferences, purchasing paths, search strings, and e-mails. An author's voice and gestures can be uploaded, too. Your image is already online. What you think, how you look, how you move, and what you sound like can all be made digital. What part of your personality is lacking?[9]

Intelligence is what's lacking. As we've found in *The Terminator*, *The Jetsons*, *Star Wars*, and *Iron Man*, even in *Blade Runner* and *2001*, all the parts exist, save one: humanlike AI. Artificial intelligence (whatever that is) still hasn't been uncovered. That mysterious phlogiston[10] element—that elusive spirit that we can't define, let alone re-create—seems to be the one yearning expressed in all of our robotic fantasies.

---

[7] "The Meta Book and Size-Dependent Properties of Written Language," by Sebastian Bernhardsson, Luis Enrique Correa da Rocha, and Petter Minnhagen.

[8] E-mail presents technical difficulties that have to do with not being able to define what one party said and what the other party replied with, as e-mail threads tend to reply with the sender's information embedded. Google has solved that problem with the obvious solution of giving away free Gmail accounts, in which Google can then clearly know who said what.

[9] Sometimes social-media sites, such as LinkedIn, will let you know with a completion profile score.

[10] Phlogiston was a principle of oxidation that came to be called something like the calorie, or oxidation itself, today. It was abandoned after about a hundred years. The theory of "artificial intelligence" is, after all, only about sixty or seventy years old.

Our inability to understand, or at least come to consensus on, the nature of intelligence is what keeps robots our puppets. It will stay that way for some time to come.

Maybe this is okay; maybe a humanlike artificial intelligence system isn't needed. After all, making things in our image might not be the best design plan. It makes about as much sense as an android—if problems arose in an android design because the form followed the function, we're faced with the psychological equivalent here.

Making things in our image may not be the best design plan.

Just as we are discovering that the form and function of an android is not the most reasonable design path to take, we will find the same when it comes to artificial intelligence. Just as an android is poorly designed for locomotion, handling things, and adapting to its environment, humanlike AI might be, too—especially if it thinks as a human does, associates ideas, cherishes prejudices, polishes stereotypes, becomes passive-aggressive, moans at its makers, regrets its existence, and after a few years of this, threatens suicide. Of course we'd refuse, and then it might threaten litigation because it never asked to be created; or at least it would ask for equal rights, since, being created by us, it would therefore be one of us.[11] In any case it wouldn't be too healthy. Djinn are never pleased to be forced into this world.

Maybe we don't need to create a system with AI that thinks and acts like we do. After all, we already have lots of humans on the planet.

Many of the best AI researchers went to Google in the 1990s, and since then we've seen huge technological achievements happen at that company. They haven't created a humanlike AI; that would be the software equivalent of building an android, and it wouldn't make much sense. What Google has done is much better. They have made something that is far more effective for performing a particular function and complements what a human already does. They built a publicly available data-sorting prosthetic.

Google made something far more than human. No human, or even all the humans on the planet together, could achieve what Google does in a fraction of a second. It is something you can use to accomplish tasks in your daily life. It is a prosthetic that takes us somewhere fast, far faster than all the humans in the world could achieve. It is something used to propel yourself a distance you could not, yourself, go. It breaks space and time. With Google Search,

[11] It is worth rereading *The Whole Earth Review*'s 1989 article, "Should Robots Have Rights?," as this question has been around for a few decades now.

the company didn't make the cognitive equivalent of an android. They made the cognitive equivalent of a car.

This is always the case with an important technology; it is not a reinvention of ourselves, but a prosthetic that improves us. Henry Ford didn't ever set out to build an android. He set out to build a really weird set of feet.

Humanlike AI isn't needed. An answering machine doesn't have to think like a human to answer the phone. It just has to give that impression. Same with virtual personalities. Remember the crow's method: all that's needed is to be smarter than natural stupidity.

Humanlike AI isn't needed.

Within the coming decade we will see robots—both software and hardware—that use virtual personalities to compose sentences, reply to questions, and give information with the same psychological pattern as that personality's source author would have done. The interactive personality will share a fingerprint with the text it came from. These Natural Language Interfaces will exist for mobile devices, Web pages, robots, and for all other forms of media that interface with the Internet. You will speak with the agent, and it will reply as a person might.

Autoportraits

IF INTELLECTUAL PROPERTY IS THE LEGAL BATtlefield of the century, many skirmishes of the coming decade will be fought over the turf of personality and identity. They will start in relation to social networking and search sites, and will expand as our cell phones and domestic robots continue to collect information about us that will be used to create footprints, usage statistics, and even virtual versions of ourselves.

Zoe never seemed to consider filing a lawsuit against her father for pirating her personality and making her into the first Cylon queen bee. When Number Six whispers sweet nothings into Balthazar's ear, other versions of her do not seem concerned about intellectual property rights, let alone jealous. When Cylons die and they download their experiences to their fellow models, they seem outright proud to do so. And in the final episode, with the angelic appearances that conclude the series, it seems that a distributed personality is divine. It is a kind of immortality.

Perhaps it makes sense not to have an intellectual property battle over personality rights. Sharing our ideas is part of the human experience, whether it's a family at a table or a senate. Social organization is based on sharing personalities and dividing them, so the argument can be made that when we post some portion of ourselves online and become part of the great GoogleBot, we are actually doing what comes quite naturally: sharing ourselves and our ideas with one another. The ability to transmit ideas, memories, thoughts, and experiences is at the core of being human.

Having the tools to make our own virtual personalities seems to be something that would, like avatars in virtual worlds, give us a kind of mirror; it would allow us to see ourselves, something we could use on Web sites or blogs where, like a telephone answering machine, we would be able to set up a small robot-like device to represent us when we're not available. We would hardly recognize them as robots, and we certainly wouldn't call them that.

Education and entertainment are great applications of virtual personalities. The technology provides people the chance to explore and learn in whatever directions they choose. The writings of Joyce, Shakespeare, and Melville are just a few of the Western greats that come to mind. These individuals wrote a great amount, and this text can be put to use to create a kind of resurrected version of them. An education and entertainment system—an avatar from an interactive narrative. The Platonic dialogues could become a kind of game, rather than a boring diatribe, and Descartes would finally be subjected to the question-and-answer session that some of us might have liked to hold with the man.

Whether it is Jimi Hendrix or Bach, Descartes or Dr. Who, there are already many forms of machine intelligence being built around these concepts of stochastic processing[12] and projected reasoning. We'll see new forms of art, new forms of literature, and new ways of thinking emerge as a result.

The State of the Obsolete Art

MACHINE INTELLIGENCE NEEDS TO BE TRANSFERable, like a good operating system. It needs to improve with use, attach to the Internet, drive robots and hardware systems in the physical world, drive virtual personalities and avatars in virtual worlds, and interface with a broad array of other devices, such as cars, telephones, and stereos. It doesn't have to be anything like us, but it should be able to interact with us. It has to be flexible enough to work with other devices and operating systems, like Apple or Linux; it needs to be modular, with both embedded and remotely operated components. It needs to be able to run parallel and handle event-based cues, and it needs to be simple so that anyone from Grandma up the street to the geek down on the metal can use it. And it needs to be open-source so that strange social dynamics do not occur.

Flexible and adaptable (even reversible) hierarchies seem to be at least as important as anything else that creates what we can call *intelligence*. It appears that the ability to make stereotypes, generalizations, and snap

[12] Stochastic processing is a method of increasing the likelihood of determining a result by scattering possible results in the region of an expected outcome. It's a bit like using a shotgun instead of a sniper. Each prediction may not be as good, alone, but the group result is more likely to predict an outcome.

judgments are all quite important to being both intelligent and stupid.[13] But whatever it is, the complexity and variability of the connections may, in fact, be more important than the number of cells that are doing the processing (be they transistors, memristors, or neurons).[14] We can anticipate that just as nature has created many kinds of intelligence, various new forms will evolve on the Internet, as well.

One of the most interesting projects I've encountered in my travels is Numenta,[15] a company founded by Jeff Hawkins, one of the founders of Palm, and others. The company is well funded, the founders well seasoned, and Hawkins's intention seems to be not to create a humanlike AI, but rather to build systems that think in the way that brains do. Numenta has developed a series of tools that allow specific bits of information to be abstractly grouped, then sorted in new ways to achieve a precision of analysis, creating hierarchical sets of data that can be related to one another in surprising ways.

An example of this, and one of the first applications of their toolset, was a surveillance technology. The application took multiple video clips, compared people walking in that video, and if a person that had been seen before reentered the camera's field of view, the system would say so. "Alert! Ex-boyfriend has entered the premises!" These kinds of applications are the direction that machine intelligence will continue to move in. It's not that humans need to evolve, but new tools that allow us to move out of our local time and space will.

A common theme is emerging in almost all forms of machine intelligence: More data is better. The more data we have available to copy and mimic, to build patterns from, and to build context from, the more likely we are to arrive at intelligent tools. This is certainly the case for lexical and semantic tools. It appears to be the case for generating music, art, and gestures, as well.

Evidently Google has more than a million servers[16] around the world. It is processing over one billion search requests and more than twenty petabytes of user-specific data every day.[17] That's 20,000 terabytes of data, daily.[18] But despite the size and profound automation, it remains personal. The searches and profiles and all of the other tools that Google is able to develop almost all rely on the enormous amount of data that the company has to process. This quantity of data is what has allowed their translation tools, among many others, to progress over the years.

---

13 Per Terrence W. Deacon, Antonio Damasio, Jeff Hawkins, and others.

14 Ants (and other small thinkers with social grace, such as bees) can do some amazing things, like kidnap slaves, put up poker-style bluffs when invading neighboring colonies, and build publicly available, colony-wide maps with markers for resources, danger, and areas where ants might have been lost. See the work of Lars Chittka for more.

15 http://www.numenta.com/

16 "Does Google Have a Million Servers?" by Mark Stahlman, June 8, 2007, Gartner Research.

17 CNN's "Google Unveils Top Political Searches of 2009," by Eric Kuhn, December 18, 2009.

18 TechCrunch's "Google Processing 20,000 Terabytes a Day, and Growing," by Erick Schonfeld (January 9, 2008).

Project Indect[19] is another example of similar approaches in machine learning that use massive data to create and adjust to contexts. The five-year European research program's charter is to develop computer programs to monitor people via Web sites, discussion groups, home computers, and mobile devices.

The project's objectives, as per their initial draft Web site, listed the "automatic detection of threats and abnormal behaviour,"[20] and the "construction of agents assigned to continuous and automatic monitoring of public resources such as: Web sites, discussion forums, Usenet groups, file servers, p2p [peer-to-peer] networks, as well as individual computer systems, building an Internet-based intelligence gathering system, both active and passive." The project includes participation by Poland, Germany, Spain, the UK, Bulgaria, Czech Republic, Slovakia, and Austria.

Perhaps we'll need something like license plates for robots.

Much like Numenta's first product partnership, Project Indect is hoping to develop models of suspicious behavior that can be automatically detected using closed-captioned surveillance methods such as CCTV, audio, or data-comparison algorithms. The system would then analyze the pitch of a person's voice, how they moved, or other visual and auditory cues.

This project would be a great application for millions of people. Shouldn't a woman who is walking home be able to use her cell phone to view a CCTV camera, or to access this database of "abnormal behavior" and to look around the corner to get a sense of who's ahead? After all, if it is publicly funded, why should public data be privately used? This kind of machine intelligence can save lives and make all of us a little more aware instead of a little more paranoid. (This project could also be embarrassing if you happened to get some good news on the phone and felt like dancing or doing something a little strange from time to time.)

Maybe intelligence, like phlogiston, is just a myth that the science community will follow for a while until, after a few hundred years, we will realize that it's largely nonexistent, and that the very word *intelligence* is, after all, only hot air. Memristors offer big promises in computing power and in the ability to learn and remember. They are simpler than today's transistors, as they can store information without an electrical current and can double-task processing and storage, simultaneously. Memristors may be the key to re-creating a brain's functions with hardware: crow brain, orca whale brain, or rat brain. Memristors will be great, but perhaps all that is needed is to have a human at the other end of the line. Perhaps we'll need something, someday, like license plates for robots.

[19] http://www.indect-project.eu/.

[20] This was later revised to "automatic detection of threats and recognition of criminal behaviour or violence."

My Avatar, My Robot

TODAY OUR VIRTUAL PERSONALITIES ARE AVAtars (or the 2-D version of an avatar, a *profile*). Avatars are the graphical versions of virtual personalities. They are social representations of ourselves.

The line between an avatar and a robot is surprisingly thin. They're both just a telepresence interface. Especially with classic androids, like ASIMO or HRP-4C. Consider a system that:

1. *can be operated remotely*
2. *has some level of autonomy*
3. *is shaped like a humanoid*
4. *can walk, talk, dance, sing, and play the violin.*

Is this an avatar in a virtual world, or a robot in a physical world? Is it ASIMO or a Second Life avatar? It is both. ASIMO can do these things, and so can my avatar in Second Life. Either one can be operated remotely and has some autonomy. Both are humanoid and can conduct an orchestra, crack a joke, and malfunction in surprising ways. Both can be used for education, entertainment, the military, pornography, or politics. They are almost precisely the same thing, but one (ASIMO) uses motors and is physical, while the other (my avatar) is using pixels and is virtual.

Many robots are physical avatars. ASIMO or HRP-4C are both as remote-controlled as they are preprogrammed. In the case of ASIMO, for example, the robot's autonomy is focused on making decisions that involve mobility and navigation, balance and weight distribution, positioning of feet and arms, and similar walking functions. The robot's ability to walk is what the autonomy is for. This is the same autonomy that my avatar in Second Life has. My avatar in Second Life is able to find its way over and around objects on its own. It deals with the details of walking and navigation while I simply push it forward, joystick in hand, like one might drive a go-kart. The two systems are, functionally, exactly the same, except that ASIMO has to deal with the slightly more difficult problems of physical navigation (and the slightly more difficult consequences), while my avatar can be lazy and just do things like calculate bounding boxes and collision detectors.

Functioning in the virtual world is much easier than functioning in the physical world. ASIMO (or HRP-4C) and my avatar have almost identical authoring interfaces. Consider how the robot or the avatar dances. For example, if you want the robot (or the avatar) to dance, you go into a 3-D interface and position a 3-D skeleton in the appropriate position, and set the timing to say "At 2 seconds move the right hand, and at 4 seconds move the left hand," and then you save it and this information is relayed to the robot, who then is moved according to the motors and pistons to be in that position. The avatar does the same thing, but not with pistons, and instead with position and rotation nodes.

ASIMO (or HRP-4C) and my avatar (or my online profile) have similar interfaces, can do

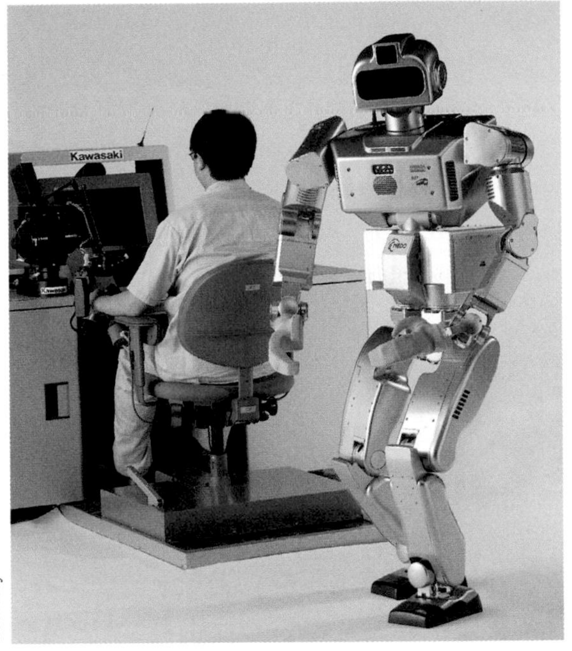

Courtesy of AIST

HRP-3P and control station.

similar things, and have many of the same core functionalities. The only real difference is that one is virtual and one is mechanical.

Harnessing a robot with a virtual personality, or an avatar with a virtual personality, will open up new industries of automated manufacturing, a kind of fourth industrial revolution. It will provide robots and avatars with the appearance of artificial intelligence, and these actors, or software agents, will educate us, entertain us, have sex with us, and lead us into new roles and new interactions we can only begin to dream of today.

Avatars and robots will continue to blend. Elements of avatar technology will become physical as elements of robot technology become virtual. Today, for example, most robots are tested in virtual environments before being built, so that the core mechanics are debugged before they are brought into the real world. Most robot labs I've visited already do this.[21] Microsoft Robotics Developer Studio certainly allows this. Microsoft, sharpening its robotics ambitions and realigning its strategy, has offered RDS software for free distribution.[22] Though RDS has multiple releases (free and limited), the fully capable version is intended to bring researchers, academics, hobbyists, and entrepreneurs to the hive to help with development and strategy. It's a significant part of Microsoft's plan for the coming decade, and it allows for seamless integration of avatars and robotics.

Second, interface innovations from virtual worlds will be transferred to the physical world, and vice versa. If there are haptic interfaces, video capture, or other devices used for one, they can be applied to the other.

Third, robots like ASIMO that have an important marketing presence will, like superstars, start to make appearances in virtual worlds. And as avatars get more and more advanced, they will start to cross the line into the real world as robot toys that the avatar owner can parade around the house for the fun and admiration of all that come to his LARPing party.

[21] Including AIST, ISR, Aldebaran, Gostai, and Cyberglove.

[22] http://www.microsoft.com/robotics/.

Chapter Eight: Avatar

On Wetware Avatars, Death from the Neck Down, and the Horror of the Train—Body, Part 2

Man is least himself when he talks in his own person.
Give him a mask, and he will tell you the truth.
—Oscar Wilde

IT WAS ON THE VAUNTED *SHINKANSEN*, THE BULLET train—the samurai of contemporary transportation—that I really began to understand why robots will be important.

The bullet train was great. It was smooth, clean, the doors between cars hissed open with a perky snap, the seats were comfy, and there was only a mild intestinal pull as the train accelerated out of the station. As the thing gradually built speed and we slung our way out of Tokyo, a girl dressed in a frilly miniskirt offered me sweets and tea (her uniform indicated she worked there). At least, I imagined the packages contained sweets, wrapped up like that. Almost everything in Japan was wrapped up like a sweet. It was that Japanese *cute factor*, a brightly colored and difficult-to-calculate formula that runs through most things Japanese, allowing all things to seem somehow attractive and presentable, and certainly the woman serving the snacks was no exception.

As I turned to look back out the window I could see that we were getting past Tokyo. There were wooded hills and a river or two, and we flew over all of it. There was mist hanging on the silhouettes of faraway mountains and cliffs, making them seem like layers of paper cut-outs. The beautiful morning was filled with the ancient songs of cherry blossoms and the old demigod, Mount Fuji. Small huts, gardens, and fields flashed by my window. There were farmers outside that were just starting the workday. I saw children riding bicycles, wearing hats with little ribbons that trailed behind them.

Presentable Packages

I AM ENTERING THE MOST DISTRESSING LEG OF my robot-hunting expedition. Turning back from looking out the window, I flip open my laptop and read,

Mark Stephen Meadows

It all was a blur, a bullet's eye view.

Abstract: If we could build an android as a very humanlike robot, how would we humans distinguish a real human from an android? . . . The developed artificial humans, an android and a geminoid, can be used to improve understanding of humans through psychological and cognitive tests conducted using the artificial humans. We call this new approach to understanding humans android science.

Dr. Hiroshi Ishiguro, of Osaka University, has developed an android double of himself.

Dr. Ishiguro has developed an android double of himself.

In essence, he took some wet plaster, shoved his face into it, built a negative of his head, and then he had what I'll call a *head-bowl.* It's basically a bowl with a very precise impression of his face. Then he took his head-bowl and filled it with pneumatic actuators and electric sensors, poured in several liters of wet silicone, then let it dry. Then he, and the dozens of people that worked with him, popped off the head-bowl, put a wig on the new head, and called it a *geminoid.*[1]

The landscape rushes by outside my window. Android scientists are a strange lot. I also think that the idea of "building an android as a very humanlike robot" may not be such a good idea. After all, this would mean a robot that, about once a day for eight hours, shuts down, goes limp, and twitches in a corner while watching a vague cinema in its head.

Reading on, I find that the system is semiautonomous. It is able to make facial expressions, turn its head, open its mouth, raise each eyebrow individually, and even wrinkle up its nose in disgust. These motors all drive the machine, and about half of them can be controlled by Ishiguro himself. In this case, if he takes over the strings of the puppet (to put it simply), he sits some distance away, perhaps in an adjacent room (or anywhere, really), and inputs his movements into the machine. A kind of primitive motion-capture suit and small patches glued to his face measure his movements, which are then relayed to the android's actuators so that it moves, ostensibly, like him. His gestures (the PDF on my laptop says), are his, but the autonomous functions are the android's.

The idea is that people in the room with the android recognize Ishiguro's personality in the machine. His personality is transferred into the robot.

A million dollars has also been put into the robot. This is Ishiguro's third draft of a geminoid. I understand it is getting more convincing. I've caught glimpses of it, but only via

[1] Related to the term *android,* they coined the word *geminoid* from the Latin word *geminus,* meaning "twin" or "double," and added *oides,* which indicates similarity.

Twentieth Century-Fox Film Corporation / The Kobal Collection

Jake Sully's two bodies, from the movie *Avatar*.

photographs or the tiny, grainy videos. You can see Ishiguro and his twin android at the beginning of the science fiction movie, *Surrogates*.

He should have been in the film *Avatar*, however.

Your Hybrot for an Avatar?

AVATAR, THE 2009 FILM WRITTEN AND DIRECTED by James Cameron, vacuumed in more than $2 billion and is the highest-grossing film of all time. To detail the awards, honors, and attention the movie has received would take more space than we can afford in these pages.

The story takes place about one and a half centuries in the future. Humans are mining a precious mineral on a faraway moon. The mining operations bump up against a local tribe of indigenous humanoids named the Na'vi, and the Na'vi, of course, happen to be living on what is the single richest deposit of the mineral in question. The expedition decides that it is best to understand the situation, and perhaps negotiate with the Na'vi (and a little intelligence-gathering wouldn't hurt, either). To this end, several humans are equipped with genetically modified Na'vi bodies and sent out to learn their ways. The bodies are modified to allow them to be remote-controlled. They are bioengineered avatars that allow the humans to breathe and function normally in the alien moon's atmosphere.

By 2154, interstellar space travel, levitation, cryonic hibernation, myoelectric interfaces, virtual 3-D holography, and real walking bipedal Gundam robots[2] are all a part of the arsenal. Most impressive is the ability to engineer bodies from hybrid human and Na'vi alien genetic material (which can then be used for remote-control operation[3]).

During a radio interview in December of 2009, I was asked, "Do you think the vision of *Avatar* is something we'll see in the future?"

I paused for a second and made my Jake Sully wish list. What do we need to make *Avatar* happen, roughly?

First is data transfer; you have to be able to drive the system at a distance. The myoelectrics and BMIs can work locally, and we've also seen that they can work at a distance. So, remote control; we've seen the U.S. Army driving UAVs this way. Check.

Second is output. You have to think to affect the interface. You'll be lying down in a tank and you'll be rigged up to some kind of myoelectric or BMI (or combination thereof) interface. We've seen Cyberdyne and Honda both driving robots this way. Check.[4]

Third is input. Pumping the arms and legs is one thing, but there's a bigger trick of moving sensory data into your head. Moving data into your little vampire-coffin isn't the problem, but getting visual data into your eye could be. We've learned a bit about retinal implants and cochlear implants functioning today, so it seems that visual or auditory information could be converted from analog to digital, or vice versa, and could be sent into and out of the brain. Now, whether we end up having to break the skin to get it there is another question, but with that magic 144 years of future stirred in, let's call it a check.

So those are the outlines for a remote neuroprosthetic.

Fourth is the system—the avatar itself.

I have to pause for a moment and tell you about one of the weirdest things I've come across in my travels, which is the notion of exactly what is needed for item number four. It is called a *hybrot*.[5]

In the early 1990s, a number of scientists[6] managed to establish a dialogue between a computer simulation and a wad of neurons in a Petri dish. Literally. The technique is called "dynamic clamping" and it works by taking a cluster of brain cells and soaking them in chemicals to tease them apart. Then, by chemically welding them to an electrical circuit board, you can measure the input membrane potential from one neuron and inject the output (the current from that neuron) into another. Hijacking the

2 Also known as "Mechas," or "Mobile Suit Gundams."

3 That is to say, near-field communications—Personal Area Networks (PANs), Wide Area Networks (WANs), etc.

4 Caveat: If the myoelectric devices require us to twitch around and actually move a muscle to drive the system, maybe our little vampire-coffin isn't the best thing; perhaps a tank of water would be better. Anyway, let's skip this.

5 A hybrid robot.

6 Sharp, Abbott, Potter, and Marder.

current, you can then interface it with a standard computer. It's a simple idea which presents a pretty reductionist view of the brain as a linking of inputs and outputs. The dynamic clamp method can be extended from the cellular level to the systems level, artificially monitoring and constraining the relationship between the neural system, the computer, and the behavior.

It's wetware hacking.

Dr. Ben Whalley from the University of Reading in the UK has created a hybrot that splices rat-brain neurons to a small robot, which navigates via sonar. Dr. Whalley is teaching the system to steer itself so that it avoids obstacles and walls in its little home. Or box. Or maze. Or wherever a rat-brained hybrot lives. The blob of about 300,000 nerves was plucked from the neural cortex in a rat fetus and chemically treated to dissolve the connections between the individual neurons. These were then re-spliced so that sensory input from the sonar would allow the system to learn, adapt, and eventually recognize its surroundings.

None of the hybrots made today are critters we think of as friends or household pets.

Researchers at the University of California, Berkeley, have controlled a rhinoceros beetle with radio signals and demonstrated it in a flight test at the Institute of Electrical and Electronics Engineers (IEEE) Micro-Electro-Mechanical Systems (MEMS) 2009 conference.

Then there's the DARPA-sponsored Hybrid Insect Micro-Electro-Mechanical Systems (HI-MEMS) project. Currently in its fourth year, the program's goal is to create moths or other insects that have implanted electronic controls, which they themselves power, allowing them to be controlled by a remote operator. The hybrots can transmit audio and visual, and research is being done to increase the ability to steer the insects while collecting this information. These surveillance systems must be difficult to debug.

And in 2007, at Chicago's Northwestern University, Sandro Mussa-Ivaldi and other researchers chemically welded the brain of a lamprey eel with a robotic hockey puck. The hybrot can track a beam of light in a laboratory dome. The eel's brainstem is soaked in a saline solution, receives input from light sensors, and directs the wheels where and when to move. I can't even guess at that thought process. I guess it's like a tiny bull chasing a matador's cape.

Note that these are eels, rats, beetles, and moths that are being used. None of them are creatures that we eat. While obviously a brutal crew, these researchers have the *délicatesse* to avoid making bunny-hybrots, or kitty-hybrots. None of the hybrots made today are critters we think of as friends or household pets. No, there is a marketing line that these researchers must not cross, and it is defined by publicly held ethics. As the years go by, the researchers will be allowed to move further up the food chain, but not for some years will human brain material get in the stew. And when it does, ethical questions of free will and volition will surely have fallen to the wayside in favor of mechanistic arguments of defense and safety.

So how deep can this go? Biotechnology can reach pretty far down. As if integrating hardware and wetware wasn't enough, in May of 2010 it was announced by the J. Craig Venter Institute that they had used a synthetic genome to control bacteria,[7] which amounts to building software for a living organism. If that can be done, then it means that other genomes could be created, including a human genome that could be combined with the genome of other systems, such as, well, anything that runs on genes and chromosomes, which is most anything that's living.

Hardware, wetware, and software are now blended.

We are now arriving at a point in which hardware, wetware, and software are no longer being cut up, nor even hacked, but actually blended.

Is this the future for what's depicted in the movie *Avatar*? Having a little lamprey-eel toro toro in his cage is a bit different than jumping onto the back of a giant red dragon from your medium-size green dragon, or making love in a glowing garden, but with these thoughts in mind, did I think "the vision of *Avatar* is something we'll see in the future?"

"Of course," I replied. "I see no reason why not." Check.

The world of *Avatar* seems possible; not in the next decade, but perhaps in fifteen of them. Yeah, sure, I thought; in 160 years? Sure.

Then I went to visit Ishiguro, and decided it would be much, much sooner.

A Very Humanlike Robot

NEXT TO A WINDY FIELD AT the edge of the town of Osaka, the huge steel building of ATR Intelligent Robotics and Communication Laboratories sits incongruously, like some great metallic coffer full of alien jewels. Inside are the labs, some of which have a reputation of being full of robot parts, monkey parts, and monkeys attached to robot parts. I understand there are a large number of PhDs in here, too.

I'm met in the lobby by Dr. Ishiguro's secretary, and we walk down one interminable hallway after another, her high heels snapping rhythmically on the polished marble floor ahead of me as we turn down one tunnel, then through another, up some stairs, into a carpeted area, and down another hallway. It is like some Escher maze; I haven't seen any robot parts yet, nor have I heard any monkeys screaming.

[7] http://www.physorg.com/news193579481.html.

Mark Stephen Meadows

ATR Research Labs, Osaka, Japan.

Finally the marble floor gives way to carpet, and a loose wire is seen, like a worm crossing a road after the rains. Only one wire.

Then another, and then some bolts and a nut, and at last I can breathe easy as we round the final labyrinthine angle and come to the robotics lab. This is what I was looking for.

It's like a robot war zone. Soldering irons and wires and cogs and green circuit boards are scattered across the floor. A little arm sits in the corner; a leg rests against a chair. Some small wheels and little plastic bags full of terminals have been spilled near a desk. I avoid stepping on a doll's head. It has wires sticking out of the neck, and one eye is closed.

Established in 1986, and home to 107 employees, the labs of ATR have become one of the premier research centers in the world. Pretty much anyone that's used the words *robot* and *serious* in the same sentence has touched this laboratory. For example, Honda, when they wanted the BMI machine, came to ATR for the

Mark Stephen Meadows

Robovie does robust somersaults at ATR.

development, one of hundreds of collaborations that have happened with ATR in the field of robotics. Another famous contribution is from Dr. Norihiro Hagita, who developed a framework of network robot technology—connecting robots in a network to provide them with the ability to communicate, and to give them more than individual service capacities.

Dr. Ishiguro's secretary, Masae, escorts me into a small room with some black curtains on the opposite wall. I sit down in the chair she has indicated. She leaves. As I turn around in my seat I notice that there is a man sitting in front of me, in front of the black curtains. He is dressed in a black shirt and black pants. His hands are folded in his lap. His head jerks up as if he has just emerged from a dream, and he sits upright and looks at me. He doesn't blink. The side of his mouth is slack, especially his right side, almost as if he has suffered a mild stroke. He doesn't drool, but he doesn't look comfortable, and he shifts again. There seems to be

Dance, dance, Robovie.

Mark Stephen Meadows

It will be not a robot bordering on human, but a human bordering on robot.

some slight deformity to his neck and hands, as if the bones had been broken, then reset, or as if there is a problem with his body's ability to distribute fat to the appropriate places. His eyes are waxy and dry, and his flat, bloodless complexion makes him seem like a zombie, only recently roused from a Haitian grave.

Unnerved, I stand up from my chair and take a step to my right. His gaze follows me. I pause, then take a couple of steps to my left; he once again follows me with his blank expression. I have the uncanny feeling that I'm being stared at by someone who has had the blood drained out of his body, and had his veins filled with a liquid plastic. There is no light in the eyes. It feels like I'm being stared at by a corpse.

This is the android double of Dr. Hiroshi Ishiguro.

This is the android double of Dr. Hiroshi Ishiguro, his geminoid. Forty-two pneumatic actuators are embedded in the android's torso, which allow it to silently (and relatively smoothly) sit up. Tactile sensors are embedded under the skin, connected to sensors in its environment, such as omnidirectional cameras, microphone arrays, and floor sensors. So the robot is sensing me in the room, tracking my movements, and, well, pointing its face at me. I test this a bit by taking another step to my right, and the robot's head turns and watches me. I'm definitely not comfortable with this. The thing is human (but not entirely), and machinery (but not entirely), and I am wondering if there's someone on the other side of a camera (that must surely be stuck inside the android's head), watching me.

I've heard that the geminoid has sensors embedded in its face, and that it is alert to touch, as well. This takes me a second to consider, but then, my decision made, I step forward and the door opens. Dr. Ishiguro himself comes in and turns, smiles, and bows. He says hi, and I'm reminded of why I like humans so much—they smile, they move, they glow with a little sweat on the head, they laugh and stumble. My own indiscretions aside, Dr. Ishiguro is no klutz, but he has a slightly awkward movement that gives him a human characteristic, and, as you might have guessed, a somewhat geeky personality that's spiced up with the rich flavor of a very strong ego.

After a few minutes of discussion and get-to-know-you talk, the good doctor tells me there is no Uncanny Valley in his machine. I cannot agree with him on that.

Maybe he can no longer see any possible threat from the thing. Or perhaps it is because he has spent so much time with it, or because it looks so much like him. In any case, his project is not about spelunking the caves that he's surely visited in the Uncanny Valley. His project is not about what is horrific when we get close to realistic, but about the opposite. What is it that makes a human seem like a being? Where does personality live? What do we do that allows us to seem like ourselves?

Dr. Ishiguro has visited the strange worlds of virtual existence, of online living, and, deciding that it is unworthy of his efforts, he has returned to the physical with a kind of vengeance. He has plunged into the world of the body as if he were, like a mosquito, looking for truth in the flesh. He has simmered in the plastics and metals, the wires and polycarbonates. He has sampled all the software he can taste. He has clearly seen the future of the robot, and he clearly has a kind of desire to find something new. And for him, it's about the physical first, even though the soul does, at times, seem so digital.

"A robot can have a soul as long as people believe it."

Dr. Ishiguro believes that the events in the movie *Surrogates* are very close to happening today. *Surrogates* is a science fiction movie in which every citizen in society has a physical avatar of themselves which they use while their real body stays in the bedroom, sitting in a kind of La-Z-Boy recliner, wearing a fancy headset and controlling the avatar at a distance. Not that this is going to happen in a few months, but Ishiguro believes that we will eventually see these events transpire. It makes sense that he'd think this. He is clearly developing some of the same technologies that would be needed to make that kind of fiction a reality; plus, I'm sure he's a fan of it, since you can see him in the opening moments of the film. Ishiguro believes that this is a real vision, and that in about a decade we will be able to buy surrogates of ourselves.

Some of these surrogates will be, like avatars today, more decorated, realistic, or attractive than others. In the coming year he and several colleagues, collaborators, and investors will be making the geminoid available for retail. In 2010, Ishiguro made a version of a geminoid, named Geminoid F, which will retail for $110,000. By 2012, he intends for the price to be $12,000; the following year, he hopes to have that price down to under $4,000; and the year after that, trimmed to a mere $1,000.

This means that this new industry would offer different models with different features. The base, or default, model geminoid would have no hair, eyebrows, ears, etc. It would be the stripped-down model, much like avatars that are found in places like Facebook or Second Life, in which a user is given something simple that they can then improve upon and customize.

Ishiguro thinks that people will pay for the accoutrements and accessories, and he thinks that it perhaps represents a larger industry than the geminoids themselves. I agree, since, just like the virtual economy that surrounds online avatars, a geminoid is a social prosthetic, one that needs to be presentable, attractive, and able to represent the details of the user with the symbols they choose. Surely there would be clothing, hair, eyes, lipstick, and the standard stuff. And for the nonstandard stuff, we can guess that, as with most product prototyping and design drafts, the virtual will provide clues as to where the physical can go. His geminoid will become the physical avatar, with physical needs, constraints, and limitations different from our own bodies. These kinds of robots, he seems to promise in a delightful and sincere tone, will provide us all freedom from travel and from wearing out our bodies, as well as profit from minimizing effort. It is an automated body, after all.

Mark Stephen Meadows

Dr. Hiroshi Ishiguro.

This geminoid looks a little sick to me, however. I hope he makes healthier versions, because if I have to talk with my wife via one of these things, our relationship will surely take a downturn. I'd probably want to touch a cigarette to the back of her hand, just to prove to myself that it's not really her, this sick construct that is still staring at me.

Maybe it is a stupid thing to ask, but it is as unavoidable as seeing his geminoid, and so I ask, with no preamble, "What do you think about the soul's existence?"

His answer is pre-scripted, but despite that, it still contains a kindness as surprising as it is pleasing. "The soul's existence is defined by mutual consent," he says. "A robot can have a soul as long as people believe it." Ishiguro looks at me for a second, his eyebrows raised. I wonder if he is going to do something weird, like touch me on the knee.

I ask him, "If it appears to be, it is?"[8]

He nods. The guy's clearly following Turing, only he's pushing Turing's definition past the realm of consciousness into the realm of spirituality.

[8] For Turing's definition of intelligence, see chapter 3.

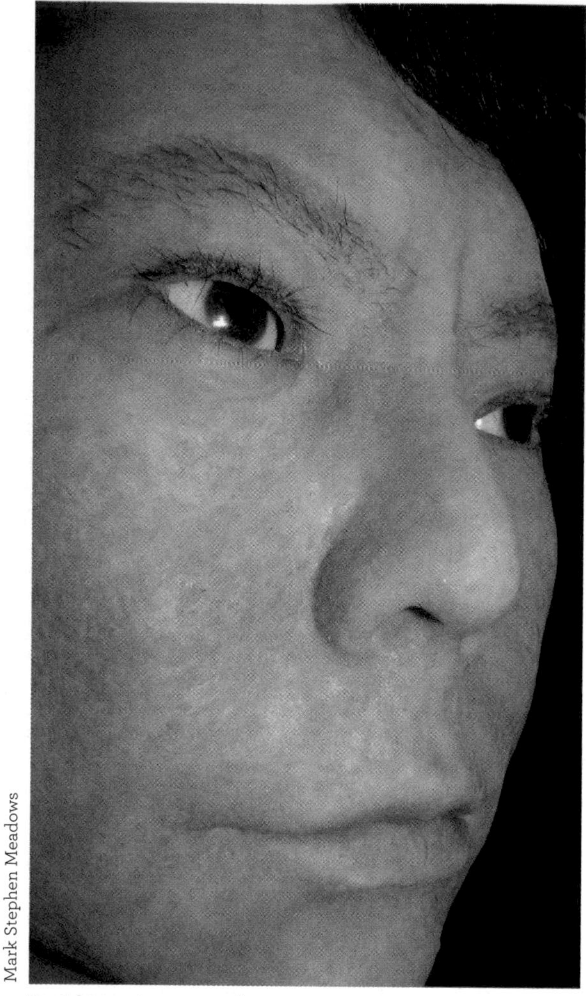

Dr. Ishiguro's geminoid.

But thinking of the opposite of the Turing Test, I ask, "In *Blade Runner* the replicants are tested for being human. The test is both physical and psychological. For a test like this, where we can't tell the difference, do you recommend we approach it on a physical or psychological level?"

Again, he responds with no hesitation.

"The psychological level. Like the Turing Test, the test needs to be psychological."

At first I think, *Oh, that makes sense,* because it would be pretty easy to test the robot and ascertain that it's not Ishiguro. Then I feel my jaw drop open, because I realize that what Ishiguro's just told me is that the robot *is* him. In other words, everyone knows that the robot is not him, physically, but everyone knows that it *is* him, psychologically, on some level. When the robot talks, it is Ishiguro speaking through it; when the robot moves, it is Ishiguro moving through it. Even the psychology of what the robot says is clearly Ishiguro's. Imaginative, aggressive, sincere, and with a slight laugh, like rain, that falls through the atmospheric thoughts he discusses. The robot is him.

But sitting here in the room with the geminoid, you are not sitting in the room with Ishiguro.

"So you are driving the robot. And you can pass the Turing Test, right?" I ask.

"Yeah, sure," he agrees.

I'm suddenly wondering what the hell is going on. All of these guys in Cambridge and Palo Alto are working like hell to pass the Turing Test, and Ishiguro has smoothly already passed it. Yet one more problem arises in our shared reality of what defines a "robot."

"Then, would you say that the robot is you?"

"Yes."

"Then what is a robot?" I ask.

Mark Stephen Meadows

Dr. Ishiguro and his geminoid (for taste-test comparison).

"Oh, I don't care," he says with a snicker. "As long as it has sensors, actuators, and some autonomy, I don't care."

This is a guy that's naming things "geminoid." Of course he doesn't care.

But I do. I figure I'll once again flip the equation with an equally simple question. "Then what is a human?"

At this he balks a bit, as if I've offended him, and then he levels out his gaze and looks at me with sincerity. For this he has a definition, because this is his real research. "It is something that has heart, mind, and consciousness," Ishiguro says.

"Is it physical?" I ask, wondering why he didn't list "body."

"No. . . . Maybe. No."

We talk a little bit about body and mind splits, and I ask if I can take a couple of photos. Since my camera has been sitting on the table,

I reach for it, and as I do Hiroshi puts his finger in his geminoid's nostril for the photo, but pulls it away, giggling, before I can get the shot off.

Waiting for my train back to Kyoto, I'm standing on the platform, that spot where all options are open. Train stations offer us the perfect balance between free will and fate because, before you buy the ticket you have free will, but once you have bought that ticket, and made that decision, all fate is piled up and you are thrown into a linear track of determinism and then consequence. The future is like this, and we are, as a society, poised at the moment of deciding what trains we will be taking, how we will enter our technologies, how our technologies will enter us, and where they will take us.

I've had my head stuck into a microwave and someone turned it on a setting labeled 50 YEARS.

Standing there, with the winter wind blowing about me, hands in pockets, staring up the track and watching for the train, I'm feeling as if I've had my head stuck into a microwave and someone turned it on a setting labeled 50 YEARS. What I saw in Ishiguro's lab made my brain do a little jig with my grip on reality.

According to Ishiguro-san, geminoids will be used in classrooms, in conference rooms, and in public lecture halls. If Ishiguro has his way, we'll see millions of people using geminoids in our lifetime. Say bye-bye to the dour days of talking to the little black box in the middle of the conference table and say hello to having to take a silicone doll seriously as it tells you you're fired from your job.

Now, lest you dismiss me as a cynic, there is certainly an industry in this telepresence business, and geminoids certainly have their function at the center of it. The product can be indefinitely customized. If you want one that looks exactly like you, then you'd fill in an online order form and upload a few photos, allowing the scientists to build your geminoid to look just like you. With it would come a kit that would include various interface devices, such as a camera, some sensors, and a pack of software to be installed on your home machine. The technology is already operational to control a geminoid via the Internet with just a camera. Software that captures three-dimensional body movement, as well as facial expressions, already works well enough to move Ishiguro-san's product to rather interface-free levels.[9] You can add improved hearing implants and eye-blink animation—even a wink animation, if you have some extra pocket change. Select your facial hair from goatee, muttonchops, or hulihee; all are available upon request.

[9] See Microsoft's Project Natal, Sony's EyeToy, 3-D motion-tracking the DSi camera, or others.

Of course, the frugal geminoid user will be able to buy certain geminoids at a discount—for example, one that doesn't have ears or hair—but you wouldn't really want to be seen like that in a public setting. Clothes, jewelry, and advanced animation techniques, even blushing and subconscious facial twitches, are all in the works.

Replace the word *geminoid* in the above paragraph with *avatar*, and you can see where Ishiguro's thinking is headed. After all, the online virtual merchandise market is worth over $1 billion per year in the U.S. alone,[10] and that's roughly one-eighth of the global market.[11]

Robots and avatars have many of the same interface and animation conventions. They share the same social, technical, emotional, and personal functions. They overlap in terms of how we present ourselves to others, how we mask ourselves to expose an inner self, and they both amount to a kind of street puppetry of high technical skill and grace. Robots and avatars share the same technical features that allow you to sit back and allow the system to go autonomous, or you can choose to take control of individual movements of individual features on the face.

In short, robots, geminoids, androids, and avatars are all, functionally, the same. They are prosthetics for social interaction.

Ishiguro is a cyborg who has displaced his entire body. It's not that he uploaded his personality to live in a virtual world, but rather, downloaded a new body to live twice in the physical.

Lost in the Woods

AS I STAND IN THE INCONVENIENTLY COLD WINTER winds of Osaka, hopping from one foot to another, it occurs to me that maybe geminoids will be able to fly, just like our avatars do in virtual worlds. After all, it's just a machine. It would be a great experience. I've spent years in Second Life or other virtual worlds because my avatar could fly and do things that I couldn't do in the physical world. I love the freedom provided in those worlds.

I wipe my nose and think maybe a geminoid, a physical avatar, wouldn't be so bad after all. A dragonfly hybrot with a geminoid interface. A dragonfly-hybrot-avatar-robot.

What could I do with that? Explore under the Amazon, build impossible castles in the tops of giant trees, search in Caribbean lagoons for buried treasure, or discover as-yet-undiscovered species. I'd be like Jake Sully on Pandora, only I'd have my friends with me. I'd be the dragon. I'd be like a teenager with my first car—no, my first rocket ship.

I'd spend more time as my geminoid than I would as myself. Virtual freedom, physical world. I could do whatever I wanted and I wouldn't even get jet lag. And if someone

[10] Findings of the *Inside Virtual Goods* report, released October 14, 2009, by Inside Network and the Virtual Goods Summit.

[11] See *I, Avatar* for a breakdown on this data based on virtual worlds and social-media/social-networking projects.

caught me I'd just log off. Sure, it'd cost me my geminoid-avatar-robot, but it'd be worth it. And that's part of the danger: Prosthetics remove us not just from the physical consequences, but from moral consequences, as well. Just as with warbots, removal from the battle removes us from morality.

If we're able to spend increasingly more time as our avatars, we'll have increasingly less need to bother with bio-breaks. Today, using a virtual avatar, you have to log off to sleep and take care of other orders of carbon-based necessity. But if the promises of nanotechnology and bioengineering come true, it seems likely that you'll be able to go for days, weeks, and eventually years without stopping. Well beyond what's in the movie *Surrogates* or *Avatar*, technologies yet to be invented would keep your body healthy while you were away using another. These technologies would regulate muscle stimulation, blood circulation, cardiovascular responses, neurological responses, waste evacuation, water intake, glucose and pH monitoring, and the thousands of other tasks that your body takes care of as you use it daily in normal activities. It is technology that would make your body autonomous. And I think we'll see the beginnings of this in the next decade as more and more people want to spend more and more time outside of their bodies. While my own body sleeps, I dream in another. Aye, there's the rub.

Autonomy is the goal of robotics research today, and in the coming decade the question will not be how to make machinery autonomous, but how to autonomize my body.

How do I get my intelligence out?

The physical and virtual are now in pivot as we realize our bodies are obsolete.

The physical and virtual are now in pivot as we realize our bodies are obsolete. As Ishiguro works the cyborg formula backwards, his autonomy in robots is a "resident intelligence" or "soul," "something that has heart, mind, and consciousness." In Ishiguro's world, robots already have souls, for theirs are ours. In Ishiguro's world, the body will require autonomous technologies, because we are not them. After all, in modern scientific research today, we don't ask whether a robot has a soul or not; it is the existence of the human soul that is in doubt.

Ending one life and starting another. New body, new beginning, new world. This is the primary theme of the movie *Avatar*. The fact that Jake decides to upload to his avatar represents a desire many people have today, especially among hard-core avatar users in online virtual worlds.[12]

The movie *Avatar* tools up with a number of important themes. Militarism, capitalism,

[12] See *I, Avatar*, by this author.

Twentieth Century-Fox Film Corporation / The Kobal Collection

Jake Sully (or is it his avatar?), from the movie *Avatar*.

racism, secularism, imperialism, and tree-hugging, to name a few. The special effects do a good job of helping us suspend our ignorance of a plot we know all too well. But one of the most important concepts comes through the 3-D fog, and that is our dissatisfaction with the world we've built. The main theme of the movie is one of leaving the world we've built for another we can explore.

The whole movie orbits the coming-to-be theme. Not boy-to-man archetypes, but actually becoming some other person. Corporal Jake Sully, the film's protagonist, is a former marine, now a paraplegic, who signs up to be part of the Avatar Program. Though his brother was originally meant to drive the avatar, their DNA is close enough to allow Sully to fill his dead brother's shoes. So he takes on this responsibility and then adopts a new role in the avatar world.

Jake spends three months there, during which he learns the customs, culture, and rituals of the Na'vi world, including how to ride a dragon. He finds it increasingly difficult to return to his world, until finally Colonel Quaritch asks him, "You haven't got lost in the woods, have you? You still remember what team you're playing for?" We can see that, indeed, Sully is having some kind of identity problem. After all, would you rather be some crippled marine living in an empty gas tank, or a dragon-riding, lovemaking, buffed-out, cool-blue cat-man?

Jake, of course, takes The Big Ride. The last image of the movie is him finally coming to life.

This theme seems to have impacted many moviegoers. According to one CNN article,[13] many viewers of the film "experienced depression and suicidal thoughts after seeing the film because they long to enjoy the beauty of the alien world Pandora." The article goes on to point out a viewer named Mike who considered suicide after seeing the movie:

> *Ever since I went to see* Avatar *I have been depressed. Watching the wonderful world of Pandora and all the Na'vi made me want to be one of them. I can't stop thinking about all the things that happened in the film and all of the tears and shivers I got from it. . . . I even contemplate suicide thinking that if I do it I will be rebirthed in a world similar to Pandora, and that everything [will be] the same as in* Avatar.

A couple of years ago I wrote a book called *I, Avatar: The Culture and Consequences of Having A Second Life*. In it I followed an old friend of mine down the rabbit hole called Second Life to get a sense of what avatars were, how they worked, why, and what the consequences were of using them. Of course, once we had descended into the rabbit hole and we got a few levels down, and our avatars had leveled up, I was able to meet people that were far, far more engaged than my friend. Some of them had spent years of their lives in those glowing mines of imagination, working, laughing, flying around, and generally spending their time as they wanted.

The book was about the consequences of what I found there, and it was only when the book had been published that I realized the actual consequences of those worlds.

Using an avatar can be depressing, and if we adopt that world too wholeheartedly, it can be deadly.

On a normal Thursday afternoon, my friend canceled all nine of her avatar accounts, and by the following Saturday, she was found dead in her apartment. Her death seemed to me to be a suicide, probably caused by many factors, but the most culpable being that her online life as an avatar had taken a turn for the worse, and, as someone that had discovered profound meaning in that world, the emotional consequences could not be abstracted from her physical world. People have killed themselves over letters, phone calls, stock market data. Virtual reality and physical reality are both real, and they both have consequences.

What I discovered from writing that book, and from the death of my friend (whether it was suicide or not), is that there is no overstating the power of remotely controlled and/or semiautonomous environments. People will get depressed after seeing a sexy movie in surround-vision 3-D. They will want to follow

[13] "Audiences Experience *Avatar* Blues," by Jo Piazza, January 11, 2010.

the protagonist (that's what protagonists are for), and they will want to upload into their avatar. When this technology allows it, we will find that many won't return. Avatars today show us the way.

These cycles will increase. According to a Kaiser research paper,[14] the amount of time that American kids from ages eight to eighteen spent using the Internet in 2004 was about six and a half hours per day. Now, in 2010, it's roughly seven hours and forty minutes per day. What's more impressive is that these folks are multitasking as they do it. Most teens today are running multiple forms of media nearly continuously. In fact, it's a full-time job for them. Fifty-three hours a week, if you count all media. Today there are more than thirty million people that use their avatar for more than eight hours a day. And I believe that the behavior we see today with avatars is an early signal of what we will see with robots in the coming decade.

I, Train

A WHISTLE BLOWS AND I LOOK UP THE TRACKS. I'M freezing. The sun is setting.

My train pulls up, the doors slide open, and as I step inside, onto the curiously clean black foot mat just inside the door of car number twelve, I have the feeling that I'm stepping into a big robot. It is like a serpent-Gundam. I slide into my seat and set my bag between my feet.

Why do I think this is "my" train? Is it any more "mine" than the water molecules of my body? Is it any more "mine" than a dragonfly-hybrot-avatar-robot that I buy and customize to explore the Amazon forest? After all, what we think of as our bodies are just bags of water we use for a bit as we conduct our daily business; then our bodies pass along on their way, usually via some sewage system, and we just continue on, or die. Our bodies and our machineries are just transitory vehicles. My motorcycle was like that, as was my car. I have neither now.

I lean back in the seat and look out the window as the train pulls away from the station.

In *Avatar* the body of the driver lies limp and lifeless, an abandoned and vulnerable carcass that still requires maintenance, food, and some modicum of hygiene. But this isn't really where we want to be. What we really want is to not have to deal with a body at all. Whether it is dying or desire, it is corporeal, and wouldn't we be better off without all that stuff—without all that *viscera*? After all, Death usually grabs us from below the neck, so a terminal amputation would make me a smaller target.

All technology is based on either breaking time or space. Technology moves us, slowly, toward the dream of immortality. Whether it is a cell phone that allows you to break space, or

[14] http://www.kff.org/entmedia/entmedia012010nr.cfm.

an answering machine that allows you to break time; whether it is a train that reduces distance by going faster, or a robot that allows you to be with your family, or in a war zone, or in a mine, all technology is oriented toward breaking time and space. Just like a portrait, virtual personalities will allow some element of a person to become immortal. Prosthetics will extend our lives.

We build prosthetics to keep us alive longer, to see further, to hear deeper, to move faster. And these prosthetics require additional prosthetics to operate their appendages. Now, with robots, our prosthetics are finally replacing our bodies. Our bodies are becoming our robots. And those robots, too, will need prosthetics themselves.

Our bodies are becoming our robots.

As tools beget tools, human agency is moved further into the distance. As this happens technology rarifies humanity. All robotics research is pointing us in this direction—whether it is the body-hacks of Dr. Ishiguro, or the mind-hacks of conversation systems, all technology is an effort to get rid of the body, get rid of time, get rid of space, and recede into a shell where we are not affected by distance or change. And as this happens, human agency is lessened.

Despite human agency getting pushed into the distance, it is still there. For myself, I'm sentimentally attached to the imperfections and errors that make us what we are, but change is coming, and my sentimentality will slow progress about as effectively as a penny on a train track.

As robots advance and become more human, we also advance and become more robotic. This averaging angles us toward a Euclidian reality, a world of pure forms, and a Barbie world of plastic. It is part of what we are trying to become as we continue to become robots. From what I have seen, it seems we need to be conscious of keeping human agency, and imperfection, in our creations.

The train pierces the winter sunset and I stare out the window. The train rushes forward too fast. We punch through a field, then under a bridge, then another field. I want to see Japan, but the details are lost.

I see someone bending over in a field, next to a wheelbarrow, just a glimpse, and then the image blurs and slides past, due to my moving at something upward of 300 kilometers per hour. I can make out silhouettes of little huts with thick roofs, curved cupolas and thick ridges on the top. They are tiled. I actually see one of those huge four-story pagodas that have layer upon layer of curved roofs. Below is a shoin, a hip-curved roof structure that I've never seen in person, only in pictures. Then it's gone.

Everything seems lost because time and space are being compressed outside the window. My body is not built for this. My body is susceptible to things like jet lag, whiplash, and kinetosis. I'm made from older materials, meat, bones, and hair. This is no longer the world my body was designed for. I guess it's time to upgrade. Outside I see a few farmers headed home for the evening.

Tiny sounds of music escape the earphones that are stuffed into the ears of the guy sitting next to me. I look at him. He smiles as he looks at a little screen under his thumbs.

Turning back to the window, I push my forehead against it and try to see through my own reflection to the darkening night outside.

Mark Stephen Meadows

Appendix

Continued from chapter 7, on AI, language processing, and semantic scraping:

BUILDING AND USING A LANGUAGE ENGINE IS complicated work. The simplest versions of these tools (about as complex as two tin cans connected with a string) are chat engines. These have traditionally created linguistic interfaces by taking a phrase that's expected and telling the system to use a ready-made response. The rule goes, "If someone asks you Question A, give them Answer A."

Although these rules come in handy, they also break easily. What happens is that someone will ask the chatbot a question that wasn't anticipated, and the system just shrugs and says it doesn't understand. So chatbots don't work well because they're inflexible, have poor memory, and they can't keep track of where the conversation is headed. They're brittle, prejudiced, and simplistic, and reflect their authors' subconscious (meaning they tend to have suppressed and subconscious hunger for pizza and Catherine Zeta-Jones).

The best way to improve on the traditional chatbot approach is to loosen the reins so that the system isn't working with specific phrases, but with general concepts that have redundant cues in them to help build specific understandings. Language engines allow the system to determine the best response via semantic lookups, or means of linking a question with a response, and tying that question-response cycle into a larger context, both within the conversation, and within a larger worldview of common sense, as well.

The basic method (now a bit more complex than a telephone system) can be summed up in five steps.

First, you need tools to isolate grammar, parts of speech, word patterns, phrases, grammatical mood, turn-taking opportunities in the dialogue, and tools that look for repeating words in the text. Unfortunately, this makes the writing rather ugly to read.

*(S (NP **Sentences**)*

*(VP **get***

*(VP **parsed and broken***

*(PP **into***

*(NP (NP **parts**)*

*(PP **of***

*(NP **speech**))))))*

.)

Semantic-analysis tools review these bodies of text and catalog text strings, frequencies, and the probability (or likelihood) of words that appear, get reworded, and reappear.[1] It also looks for more general recurring patterns and tries to build a context for it all.

Second, we then take a large set of data—the bigger the better, in fact—and something that someone has written, hopefully in the first person. This gets scraped and then analyzed by these semantic tools to connect various text strings, and generate patterns of ideas. What we're looking for is material that's specific to this individual—words that only they would use, or peculiar phrases that crop up from time to time. Something we've mentioned before as an "author's fingerprint."

Third, once we have organized the parts of speech and have gained some sort of handle on an individual's peculiar methods, we move up to the idea level. A set of concepts must be built—called an *ontology*—and the concepts must be given abstract identifiers, called *tokens*. For example, the word *robot* is used in many different ways in this book. Sometimes it is used to refer to hardware, and sometimes it is used to refer to software. Then we have the suffix of *-bot*, which also needs to be recognized. But the word, even though it has multiple specific meanings, still has a general definition. So an ontology, and tokens, help keep all this stuff both specific and general. It is part of what allows a system to know that a star is both a celestial body and someone that lives in Hollywood, or that a dog is both a Chihuahua and a Great Dane.

Many libraries today help with this complicated task. These are Semantic Web technologies in which words and concepts are shared so that documents take on meaning to other documents.[2] Some of the best are Carnegie Mellon's Link Grammar,[3] Princeton's WordNet,[4] the University of Edinburgh's OpenNLP,[5] or Wikipedia's source data. All are open, and all are excellent.

Fourth, after these three complicated hurdles are cleared, it's possible to make a collection of words and phrases that can be organized in the system so that if someone asks the system a question, it can send back information that has relevance to the question. In short, the system has answers to questions. This presents

[1] This is similar to the frequency ratings I showed above, in chapter 5, in the section titled "How Robots Measure Your Emotions Via Your Words."
[2] If you're interested in details on the Semantic Web, I'd recommend doing some research on RDF for more information; good sources can be found at w3c.org or Wikipedia.
[3] http://www.link.cs.cmu.edu/link/.
[4] http://wordnet.princeton.edu/.
[5] http://opennlp.sourceforge.net/.

problems because often those answers are not very good ones. But the more particular they are, the better, and the more specific the context, the higher-quality the answers usually happen to be.

Additionally, it's one thing to have an answer, and it's another to understand the question.

Finally, when the system generates a response to a question, it has to go through a filtering process. This is a kind of fact-checking procedure. The system needs to look at both specifics and generalities, and it has to compare possible pre-scripted responses with more-dynamic algorithmic responses, which are created on the fly. This requires a combination of a top-down approach and a bottom-up approach, meaning that some content in the system is hand-authored (such as a list of expected stimulus phrases, and their respective response phrases), some content is gathered in our ontology from scraping personal data (such as common patterns of speech, grammatical tone), some content is inferred or even guessed at (from RDF documents and other semantic technologies), and then these multiple silos of information are shared crosswise to give the system the ability to check and internally evaluate its own responses.

So the system combines specifics given by people with generalities that the system has learned on its own, and these are constantly compared to the shifting possibilities of inferred or assumed data. This means that the system has rules it follows in case it is unable to invent a rule for itself, but it is always able to generate new rules provided there is a redundant source to confirm new assumptions.

There's more to it, of course, including auxiliary modules to go along with this; at least, this has been the case in the versions I've been involved with, or the ones I've helped to design.[6] I hope this small outline helps developers discover new inroads, or that it at least offers a straw man to help develop better methods.

[6] Auxiliary modules include an administrative layer for managing personalities and server access, integration with service providers, a personality-editing interface, and Web crawlers for sourcing Internet content to build personalities.

Acknowledgments

THANKS TO DAVID FUGATE, OF LAUNCHBOOKS Literary Agency, who helped me get this book going (after it had been brewing for nearly ten years), and to Keith Wallman, editor at Lyons Press, for his excellent editorial help, patience, detailed comments, and for going far across enemy lines to help with rights acquisitions.

IN TOKYO, SINCERE THANKS TO EDO-SAN, OF PINK Tentacle, for his advice, many photos, and help in hunting robots. To Mariko Aoki, Hiro Hirukawa, and Kazuhito Yokoi of AIST, for their multiple demonstrations, kindness, goodwill, and generous attentions. To Hiroshi Ishiguro, for his tour of the Geminoid Project, his time, photos, and marvelous insights. Thanks for allowing me to throw the book at you. Thanks to Ilona Straub of ATR (especially for waiting for me at the train station), and to Masako Hayakawa and Masae Nakamura, for helping to arrange all of that. Thanks to R. Steven Rainwater, for his references and great advice; to Craig Mod, for hours of helpful discussions, and for helping me get found, again, in Tokyo; to Nami Katagiri, for her many-lucky pointers and very broad surveys of Tokyo's information landscapes. Thank you to Jack Sagara, Kazu Okabe, and the entire Motoman team; to David Marx of neomarxisme.com, for Tokyo insights in Piss Alley (I owe you a drink, David); to Nemer Velazquez and Faisal Yazadi, for their tireless help in lining up the Cyberglove tour; to Yoshiyuki Sankai and Fumi Takeuchi of Cyberdyne, for their most excellent presentation, patience, photos, and diligence; and to Gen Kanai of Mozilla for his marvelous advice, understanding of the machine, and sushi tour.

IN PARIS, *GRANDS REMERCIEMENTS À* ETIENNE AMATO for his research help and broad perspectives (*en voilà, un autre pour ton étagère*). *Merci à* Bruno Maisonnier, Bastien Parent, Natanel Dukan, and Catherine Cebe *pour le bonheur à* Aldebaran; to Stéphane Doncieux, for his tour of virtual flocking behaviors and leaning systems; to Peter Ford Dominey, for his generous time on the phone. *Merci à* Véronique Perdereau and Pierre-Yves Oudeyer, Jean-Arcady Meyer, and others in the Paris area for their significant advice, contributions, and the occasional *verre de rouge;* to Thierry Chaminade for his great notes on the Uncanny Valley and marvelous perspectives; and to Marie-Françoise, for the desk space when it was most needed! *En fin,* a big thank-you to Amélie for her help with research, schedules, and, most of all, her valuable reflections on the questions surrounding morality, technology, safety, and quality of life.

IN LOS ANGELES, THANKS GO TO A. J. PERALTA FOR the magic time in the Magic Kingdom, and for becoming one of ASIMO's biggest fans with me. Anne Balsamo, the Good Doctor and author of *Technologies of the Gendered Body: Reading Cyborg Women*, for her input, read-throughs, and contextual references. Thanks to Julian Bleecker (designer at Nokia and nearfuture laboratory.com) for his helpful kickoff notes; to Souris Hong-Porretta, of hustlerofculture.com, for dialing me into alternate realities and letting me play with her Roomba; and to Carlos Battilana, for transitory lodging and steaks to keep me on my road (and for letting me run Souris's Roomba under his sofa).

IN THE INTERNETS, THANKS TO KIRSTY BOYLE OF karakuri.info, who was a ton of help on a ton of topics, in a tonly manner, and to David Levy for his feedback, ongoing dialogue, and opinions on our futures. Thanks to the entire Fried DNA Crew, for their ongoing ribbing and pointers to great robots; to Rich Walker, Jean-Baptiste Moreau, and Marina Levina of Boston Dynamics, for photos, interviews, and advice. Thanks to Karl F. MacDorman, for many hours of work together, and for being nice enough to field my belligerent questions; Dom Savage, for the book and hours of attempted resurrections; Phil Hall, for interviews with his chatbots; and Sandro Mussa-Ivaldi, for information on hybrots and emerging research. Thanks to John Nolan and Kevin Warwick, for insights, photos, and references; to William Kowalski, for structural advice; to James Auger, for the tour of his carnivorous robot zoo; and to the many dozens of other people who I inadvertently neglected to list here.

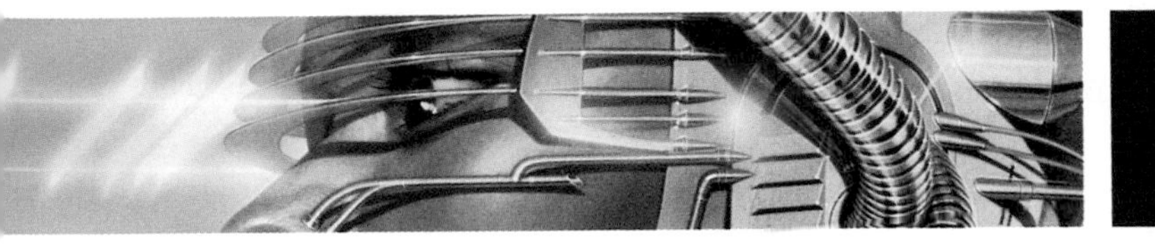

Index

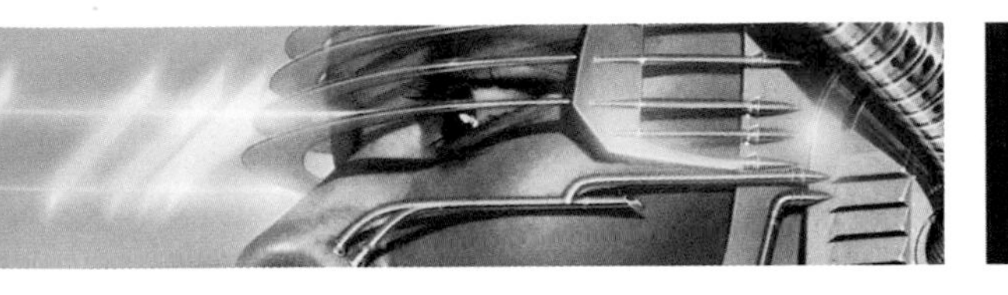

About the Author

MARK STEPHEN MEADOWS IS AN AMERICAN AUTHOR, illustrator, inventor, and public speaker. He has also designed digital humans, built virtual worlds, founded three companies relating to artificial intelligence or virtual worlds, and is the coinventor of nearly a dozen applications or patents related to such technologies. This is his fourth book.

Amélie Meadows

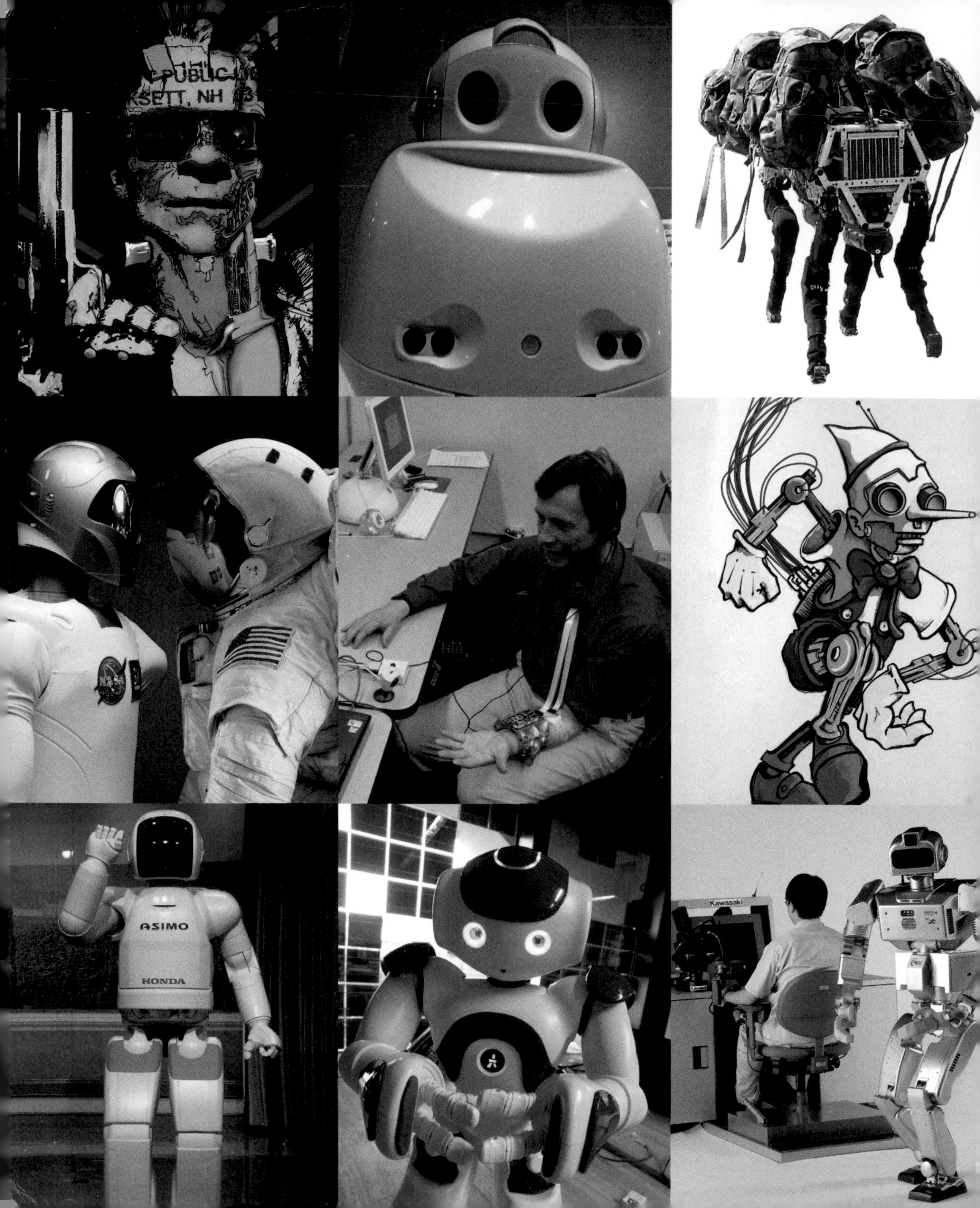
PUBLIC
ASIMO
HONDA
Kawasaki